SpringerBriefs in Space Life Sciences

Series editors

Günter Ruyters
Markus Braun
Space Administration
German Aerospace Center (DLR)
Bonn, Germany

The extraordinary conditions of space, especially microgravity, are utilized for research in various disciplines of space life sciences. This research that should unravel – above all – the role of gravity for the origin, evolution, and future of life as well as for the development and orientation of organisms up to humans, has only become possible with the advent of (human) spaceflight some 50 years ago. Today, the focus in space life sciences is 1) on the acquisition of knowledge that leads to answers to fundamental scientific questions in gravitational and astrobiology, human physiology and operational medicine as well as 2) on generating applications based upon the results of space experiments and new developments e.g. in non-invasive medical diagnostics for the benefit of humans on Earth. The idea behind this series is to reach not only space experts, but also and above all scientists from various biological, biotechnological and medical fields, who can make use of the results found in space for their own research. SpringerBriefs in Space Life Sciences addresses professors, students and undergraduates in biology, biotechnology and human physiology, medical doctors, and laymen interested in space research. The Series is initiated and supervised by Dr. Günter Ruyters and Dr. Markus Braun from the German Aerospace Center (DLR). Since the German Space Life Sciences Program celebrated its 40th anniversary in 2012, it seemed an appropriate time to start summarizing – with the help of scientific experts from the various areas - the achievements of the program from the point of view of the German Aerospace Center (DLR) especially in its role as German Space Administration that defines and implements the space activities on behalf of the German government.

More information about this series at http://www.springer.com/series/11849

Markus Braun • Maik Böhmer •
Donat-Peter Häder • Ruth Hemmersbach •
Klaus Palme

Gravitational Biology I

Gravity Sensing and Graviorientation
in Microorganisms and Plants

 Springer

Markus Braun
Space Administration
German Aerospace Center (DLR)
Bonn, Germany

Maik Böhmer
Institute for Molecular Biosciences
Johann Wolfgang Goethe University
Frankfurt am Main, Germany

Donat-Peter Häder
Emeritus from Friedrich-Alexander
University
Erlangen Nürnberg, Germany

Ruth Hemmersbach
Institute of Aerospace Medicine
German Aerospace Center (DLR)
Cologne, Germany

Klaus Palme
Institute of Biology II/Molecular
Plant Physiology, Faculty of
Biology, BIOSS Centre for
Biological Signaling Studies
University of Freiburg
Freiburg, Germany

ISSN 2196-5560   ISSN 2196-5579   (electronic)
SpringerBriefs in Space Life Sciences
ISBN 978-3-319-93893-6   ISBN 978-3-319-93894-3   (eBook)
https://doi.org/10.1007/978-3-319-93894-3

Library of Congress Control Number: 2018948369

Printed on acid-free paper

This Springer imprint is published by the registered company Springer International Publishing AG part of
Springer Nature.
The registered company address is: Gewerbestrasse 11, 6330 Cham, Switzerland

Foreword

Nearly 150 years ago, naturalists like Darwin, Pfeffer, and Sachs were already fascinated by plant gravity-sensing and gravity responses. At that time, they tried to study these effects by modifying the gravity level and randomizing the gravity vector with different methods and sophisticated tools like centrifuges and clinostats. Only with the advent of (human) spaceflight some 70 years ago, however, plants could be intensively investigated in different kinds of spacecrafts providing a free-fall situation—a research environment without gravitational accelerations: capsules falling in the vacuum of a drop tower, parabolic airplane flights, rockets, satellites, and space stations in the low-Earth orbit—these are the exciting high-technology workplaces and almost stimulus-free microgravity laboratory environments gravitational biologists use to clarify the impact of gravity on biological processes, cells, and organisms. The diversity of ingenious sensing and response mechanisms that obviously have evolved in parallel several times in the various kingdoms of life to perceive the direction of gravity and to use this constant environmental cue for orientation is the specific topic of the eight chapters of this latest booklet *Gravitational Biology—Gravity Sensing and Graviorientation in Plants and Microorganisms* in our series *Springer Briefs on Space Life Sciences*.

The first chapter provides a short introduction into the history of gravitational biology research addressing the impact of gravity on the evolution of life on Earth, general physical principles, and evolutionary as well as physiological aspects of gravity-sensing and the specific gravity-related responses of motile microorganisms (gravitaxis) and sessile algae, moss, ferns, and higher plants (gravitropism).

A comprehensive overview of the different microgravity-simulation methods used on ground and real microgravity platforms, which are available for gravitational biology research, is given in the second chapter. Centrifugation and microgravity simulation are not only methods used for testing hardware and preparing rather complex and rare experiments in microgravity—conducted with proper controls and considering the specific side effects, they have become valuable and permanently available tools also and not least for stand-alone experiments. With the completion of the ISS, a low-Earth orbit platform for long-term microgravity studies and series of experiments complemented the spectrum of gravitational biology research opportunities.

The following chapters summarize the current knowledge of gravity-sensing and response mechanisms of various microorganisms and plants and especially review the contribution of microgravity research to this field of life sciences. Gravitaxis— the gravity-oriented swimming behavior of motile microorganisms—and the ecological importance of the capability is the topic of the third chapter. Positive and negative gravitropism of single cells is most intensively studied and is best understood in rhizoids and protonemata of characean algae. The results of experiments in microgravity which have contributed greatly to the characterization of the cytoskeletal basis of gravity sensing and the identification of components of the gravitropic signaling pathway in both cell types are described in Chap. 4, while gravitropic phenomena in fungi, mosses, and ferns, described in Chap. 5, are still more enigmatic. Gravitropism of higher plant organs is dealt with in two chapters. Whereas the authors of Chap. 6 present a short history of research on higher plant gravitropism and focus mainly on cellular aspects of the gravitropic signaling pathway, the authors of Chap. 7 report on the dramatically increasing amount of molecular data of ground and space flight experiments. Current and future results promise to unravel the molecular basis for gravity-sensing and response mechanisms in higher plants.

The final Chap. 8 introduces several bioregenerative life support systems which have been designed and developed by the German Space Administration and the European Space Agency with two major goals. On the one hand, fundamental understanding of the complex physiological interaction between different organismic components in microgravity is the prerequisite for the development of a bioregenerative, multi-species life support system that is meant to complement the currently used physicochemical life support systems sustaining humans on space missions with food, oxygen, or other essential supplements, and on the other hand, such bioregenerative systems need to be technologically optimized to be reliable and maximal efficient to save limited resources and mass not only for long-term human exploration missions beyond low-Earth orbit but also for ground applications in remote places such as deserts, polar stations, and other hostile areas on Earth.

Bonn, Germany   Markus Braun
April 2018   Günter Ruyters

Preface to the Series

The extraordinary conditions in space, especially microgravity, are utilized today not only for research in the physical and materials sciences—they especially provide a unique tool for research in various areas of the life sciences. The major goal of this research is to uncover the role of gravity with regard to the origin, evolution, and future of life, and to the development and orientation of organisms from single cells and protists up to humans. This research only became possible with the advent of manned spaceflight some 50 years ago. With the first experiment having been conducted onboard Apollo 16, the German Space Life Sciences Program celebrated its 40th anniversary in 2012—a fitting occasion for Springer and the DLR (German Aerospace Center) to take stock of the space life sciences achievements made so far.

The DLR is the Federal Republic of Germany's National Aeronautics and Space Research Center. Its extensive research and development activities in aeronautics, space, energy, transport, and security are integrated into national and international cooperative ventures. In addition to its own research, as Germany's space agency the DLR has been charged by the federal government with the task of planning and implementing the German space program. Within the current space program, approved by the German government in November 2010, the overall goal for the life sciences section is to gain scientific knowledge and to reveal new application potentials by means of research under space conditions, especially by utilizing the microgravity environment of the International Space Station (ISS).

With regard to the program's implementation, the DLR Space Administration provides the infrastructure and flight opportunities required, contracts the German space industry for the development of innovative research facilities, and provides the necessary research funding for the scientific teams at universities and other research institutes. While so-called small flight opportunities like the drop tower in Bremen, sounding rockets, and parabolic airplane flights are made available within the national program, research on the ISS is implemented in the framework of Germany's participation in the ESA Microgravity Program or through bilateral cooperations with other space agencies. Free flyers such as BION or FOTON satellites are used in cooperation with Russia. The recently started utilization of Chinese

spacecrafts like Shenzhou has further expanded Germany's spectrum of flight opportunities, and discussions about future cooperation on the planned Chinese Space
Station are currently underway.

From the very beginning in the 1970s, Germany has been the driving force for
human spaceflight as well as for related research in the life and physical sciences in
Europe. It was Germany that initiated the development of Spacelab as the European
contribution to the American Space Shuttle System, complemented by setting up a
sound national program. And today Germany continues to be the major European
contributor to the ESA programs for the ISS and its scientific utilization.

For our series, we have approached leading scientists first and foremost in
Germany, but also—since science and research are international and cooperative
endeavors—in other countries to provide us with their views and their summaries of
the accomplishments in the various fields of space life sciences research. By
presenting the current SpringerBriefs on muscle and bone physiology we start the
series with an area that is currently attracting much attention—due in no small part to
health problems such as muscle atrophy and osteoporosis in our modern aging
society. Overall, it is interesting to note that the psycho-physiological changes that
astronauts experience during their spaceflights closely resemble those of aging
people on Earth but progress at a much faster rate. Circulatory and vestibular
disorders set in immediately, muscles and bones degenerate within weeks or months,
and even the immune system is impaired. Thus, the aging process as well as certain
diseases can be studied at an accelerated pace, yielding valuable insights for the
benefit of people on Earth as well. Luckily for the astronauts: these problems slowly
disappear after their return to Earth, so that their recovery processes can also be
investigated, yielding additional valuable information.

Booklets on nutrition and metabolism, on the immune system, on vestibular and
neuroscience, on the cardiovascular and respiratory system, and on psycho-physiological human performance will follow. This separation of human physiology and
space medicine into the various research areas follows a classical division. It will
certainly become evident, however, that space medicine research pursues a highly
integrative approach, offering an example that should also be followed in terrestrial
research. The series will eventually be rounded out by booklets on gravitational and
radiation biology.

We are convinced that this series, starting with its first booklet on muscle and
bone physiology in space, will find interested readers and will contribute to the goal
of convincing the general public that research in space, especially in the life sciences,
has been and will continue to be of concrete benefit to people on Earth.

Bonn, Germany   Markus Braun
July 2014   Günter Ruyters

DLR Space Administration in Bonn-Oberkassel (DLR)

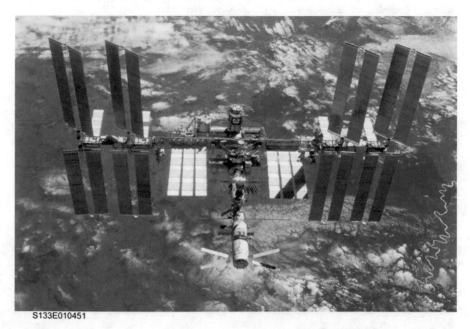

The International Space Station (ISS); photo taken by an astronaut from the space shuttle Discovery, March 7, 2011 (NASA)

Extravehicular activity (EVA) of the German ESA astronaut Hans Schlegel working on the European Columbus lab of ISS, February 13, 2008 (NASA)

Acknowledgements

Research projects in general but even more so microgravity research projects are the results of the joint efforts and dedicated work of people of multiple disciplines. We thank the numerous technicians, engineers, people from space agencies, astronauts and cosmonauts, university students, and scientists involved in the compilation of concepts, in the development and testing of experiment hardware, and in all the different phases of ground-based research projects and spaceflight missions. Scientific success of such complex missions is only possible when each step is perfectly organized and thoroughly executed like smoothly interlocking gearwheels. Unforgettable experiences and events and long-lasting friendships are our personal invaluable gains. We wish to express our sincere thanks to the teams of Airbus Defense and Space, Bremen and Friedrichshafen, OHB System AG (including former Kayser-Threde), DLR Moraba, Swedish Space Corporation SSC, NASA, ESA, Roskosmos, TsSKB Progress and the Institute for Biomedical Problems IBMP, Chinese Technology Center for Space Utilization CSU, Chinese Manned Space Agency CMSA, and Zentrum für angewandte Raumfahrttechnologie und Mikrogravitation (ZARM). In particular, we acknowledge the support of the ESA and DLR project management teams and the funding of many projects provided by DLR Space Administration (German Aerospace Center) on behalf of the Bundesministerium für Wirtschaft und Energie (BMWi).

Contents

Contributors

Maik Böhmer Institute for Molecular Biosciences, Johann Wolfgang Goethe University, Frankfurt am Main, Germany

Markus Braun Space Administration, German Aerospace Center (DLR), Bonn, Germany

Franck Ditengou Institute of Biology II, Molecular Plant Physiology, University of Freiburg, Freiburg im Breisgau, Germany

Dennis Gadalla Institute for Evolution and Biodiversity, University of Münster, Münster, Germany

Donat-Peter Häder Emeritus from Friedrich-Alexander University, Erlangen Nürnberg, Germany

Ruth Hemmersbach Institute of Aerospace Medicine, German Aerospace Center (DLR), Cologne, Germany

Klaus Palme Institute of Biology II/Molecular Plant Physiology, Faculty of Biology, BIOSS Centre for Biological Signaling Studies, University of Freiburg, Freiburg, Germany

William Teale Institute of Biology II, Molecular Plant Physiology, University of Freiburg, Freiburg im Breisgau, Germany

Chapter 1
Gravity Sensing, Graviorientation and Microgravity

Donat-Peter Häder, Markus Braun, and Ruth Hemmersbach

Abstract Gravity has constantly governed the evolution of life on Earth over the last 3.5 billion years while the magnetic field of the Earth has fluctuated over the eons, temperatures constantly change, and the light intensity undergoes seasonal and daily cycles. All forms of life are permanently exposed to gravity and it can be assumed that almost all organisms have developed sensors and respond in one way or the other to the unidirectional acceleration force. Here we summarize what is currently known about gravity sensing and response mechanisms in microorganisms, lower and higher plants starting from the historical eye-opening experiments from the nineteenth century up to today's extremely rapidly advancing cellular, molecular and biotechnological research. In addition to high-tech methods, in particular experimentation in the microgravity environment of parabolic flights and in the low Earth orbit as well as in "microgravity simulators" have considerably improved our knowledge of the fascinating sensing and response mechanisms which enable organisms to explore and exploit the environment on, above and below the surface of the Earth and which was fundamental for evolution of life on Earth.

Keywords Gravitaxis · Gravitropism · Gravireceptor · Statolith · Microgravity platforms

1.1 Introduction

"…3…2…1…Lift off!" Esrange near Kiruna in Northern Sweden—a remote European satellite station. A TEXUS sounding rocket roars out of the launch tower into the blue early morning sky. Teams of gravitational biologists and engineers follow anxiously on the video screens the separation of the first rocket stage, the ignition and burn-out of the second stage. After separation from the rocket motor 74 s after launch, the payload with its biological samples weightlessly follows a parabolic flight path up to an altitude of about 350 km before it falls back towards Earth. The 6-min microgravity phase is abruptly terminated by the reentry of the payload into the Earth' atmosphere. A parachute guarantees the safe landing of the

M. Braun et al., *Gravitational Biology I*, SpringerBriefs in Space Life Sciences,
https://doi.org/10.1007/978-3-319-93894-3_1

samples, which are then brought back via helicopter and further processed in laboratories close to the launch site.

Biology using rockets, the International Space Station, satellites and parabolic plane flights appears to be a mismatch at a first glance but provides unique and almost stimulus-free high-tech environments for gravitational biologists. Here they can study the impact of gravity on fundamental biological processes and on mechanisms almost all organisms on Earth have invented to make use of the only constant environmental cue, the gravity vector, for orientation. Changing and, most importantly, switching off the parameter of interest and studying its impact on biological processes and responses of organisms is the usual way to characterize the nature of biosensors and to unravel the molecular and cellular basis of sensing mechanisms, signaling pathways and responses. Unlike photobiologists who modify and switch off light to investigate photosynthetic processes, photoreceptors or light-induced pathways, gravitational biologists cannot simply switch off gravity on Earth. However, the influence of gravity can be randomized and strongly reduced with the aid of clinostats, 2D and 3D random positioning machines. Furthermore, the gravitational acceleration can be increased by using centrifuges; but investigations in the absence of gravity can only be achieved in so-called free-fall situations provided by drop towers, parabolic flights, sounding rockets flights, shuttles, satellites and the International Space Station in the low-Earth orbit.

The following chapters provide a comprehensive overview of the wealth of information that has been collected over the last 30 years in the areas of gravity sensing and gravity orientation found in microorganisms, lower and higher plants. The knowledge is of fundamental importance for understanding life on Earth and has significance for regenerative life support and energy systems for Earth applications and as critical components of future long-term spaceflight missions and human exploration of other planets and the universe.

1.2 Gravity and the Evolution of Life on Earth

About a century ago Albert Einstein showed the equivalence of mass and energy in the "world's most famous equation" $E = mc^2$ (Bodanis 2005). His theory of special relativity combined the electrical and magnetic forces. Modern quantum theory deals with weak and strong intermolecular forces as well as electromagnetic interactions (Margenau and Kestner 1969) but fails, 100 years after Albert Einstein, to include gravity, which is the fourth elementary force in our universe described by the theory of general relativity, in a generalized 'weltformel' (theory of everything; Hawking 2006).

The difficulty of including gravity in this theory is annoying since it rules our universe, holds planetary systems and galaxies together and traps the masses of hundreds of millions to billions of stars in the space less than our solar system in black holes which are believed to linger in the center of perhaps every of the several hundred billion galaxies which are thought to comprise our observable universe (Streeter 2016).

While the magnetic field of the Earth has fluctuated over the eons, temperatures change with time and the light intensity undergoes seasonal and daily cycles, gravity has constantly governed the evolution of life on Earth with an acceleration of 9.81 m s^{-2} (=1 g). All forms of life have been constantly exposed to gravity during the entire evolution. It controls the vertical growth of trees (Fig. 1.1) and prevents us from falling from the surface of the globe even "down under". In fact, gravity affects almost all biological, chemical and physical processes on the molecular, cellular and organismic level on Earth and is utilized in many ways by organisms since the beginning of the evolution of life more than 3.5 billion years ago.

Fig. 1.1 Trees grow vertical parallel to the gravity vector and away from the center of gravity (negative gravitropism) and not perpendicular to a sloping Earth surface

1.3   Gravity Responses of Motile Microorganisms

There has been a historical long-lasting debate as to which level of organization gravity interferes with living organisms and influence their behavior, physiology and functionality. Motile microorganisms sense and respond to a plethora of environmental physical and chemical stimuli to adjust the position in their habitat for optimal growth and reproduction (Häder et al. 2005). Prokaryotic and eukaryotic protists respond to the intensity and direction of light and show positive and/or negative phototaxis (movement toward or away from the light source) (Fraenkel and Gunn 1961; Häder 1991; Fiedler et al. 2005; Häder and Iseki 2017) as well as a dependence of their swimming speed on the light intensity (photokinesis) (Nultsch 1975; Zhenan and Shouyu 1983). Many organisms sense and respond to chemical stimuli by swimming up a gradient of attractants or down a gradient of repellents (positive and negative chemotaxis) which has been studied extensively in bacteria (Dusenbery 1985; Clegg et al. 2003; Wadhams and Armitage 2004). Flagellates and ciliates exploit oxygen gradients to control their vertical migrations in the water column (Finlay et al. 1986; Finlay and Fenchel 1986; Hemmersbach-Krause et al. 1991a; Porterfield 1997). The magnetic field of the Earth is exploited by many organisms (magnetotaxis) from bacteria to mammals (Lohmann et al. 2004; Simmons et al. 2004; Lin et al. 2017), and others detect electric fields or currents (galvanotaxis) (Votta and Jahn 1972; Kim 2013). The ambient temperature has many effects on biochemical and physiological reactions in all organisms. Therefore, it is no wonder that motile organisms utilize thermal gradients to reach habitats of favorable temperatures (Kamykowski and Zentara 1977; Kushner 1985; Steidinger and Tangen 1996; Maree et al. 1999). Some organisms have been found to be capable of detecting amazingly small temperatures on the order of less than 0.01 $°C\ cm^{-1}$ (Poff and Skokut 1977; Whitaker and Poff 1980; Fontana and Poff 1984).

Due to the ability of microorganisms to respond to a variety of environmental stimuli, the question arose whether organisms have also developed sensors for gravity and respond to the gravitational vector perpendicular to the Earth surface even if the reactions may not be visible or have not yet been revealed. Only very small prokaryotic organisms such as bacteria may not be able to respond to gravity since their motile behavior is governed by Brownian motion (Todd 2007; Li et al. 2008; Buttinoni et al. 2012). However, recent studies showing that even the fluidity of membranes is altered by gravity shed new light on this consideration and might change the view (Kohn et al. 2017). In microorganisms, movement parallel to the gravity vector of the Earth is called gravitaxis (see Chaps. 2 and 3). Initially the term "geotaxis" (Bean 1984; Fenchel and Finlay 1984) was used but replaced by "gravitaxis" since the organisms do not only respond to the gravity field of the Earth but also to that of other celestial bodies or artificial accelerations. Some organisms move up against the gravity vector (negative gravitaxis) (Hemmersbach-Krause and Häder 1990; Häder et al. 1995), others swim downward (positive gravitaxis; Bechert 2009) or are capable of doing both at different developmental phases such as the photosynthetic flagellate *Euglena* (Häder and Hemmersbach

2017). Like in the photoresponses, gravity can induce changes in the swimming velocity of a microorganisms which is termed gravikinesis (Machemer et al. 1991; Machemer 1996). Microorganisms sediment due to their higher specific density than the surrounding medium and the sedimentation velocity is vectorially added to the swimming velocity. Thus, the net swimming velocity is lower in upward direction than in downward direction. However, in some ciliates, such as *Paramecium, Didinium, Tetrahymena* and *Loxodes*, a directional-dependent speed regulation—accelerated upward swimming versus decelerated speed during downward swimming—results in compensation of sedimentation and, thus, they aggregate in a water layer with optimal environmental conditions (Hemmersbach-Krause et al. 1991b; Machemer et al. 1991; Ooya et al. 1992).

1.4   Gravity Responses of Sessile Plants

Movements of organs of sessile plants are called tropisms. Branches, fronts, stems, flowers and leaves can move with respect to light (phototropism; Briggs 2014; Liscum et al. 2014). Apical stems often bend toward the light source which is called positive phototropism (Briggs 2014; Liscum et al. 2014) while *Arabidopsis* inflorescences (flowering stems) show negative phototropism (Sato et al. 2015). With a few exceptions, roots of higher plants do not bend with respect to the light direction which in *Arabidopsis* have been found to be mediated by the universal photoreceptors in higher plants, phytochromes (Ruppel et al. 2001; Kiss et al. 2003). The young seedlings of Gramineae (coleoptiles) such as *Avena* have been studied for a long time and show a remarkable behavior: in unilateral light at low irradiances they show a positive phototropism, at higher a negative one and at an even higher one they bend again towards the light source (Buder 1920; Everett and Thimann 1968). Plants use proprioreceptors to sense their own growth. E.g., the expression of the PtaZFP2 gene is closely related with the bending angle of a poplar stem (Hamant 2013). In addition, roots of several plants have been shown to react to gradients in humidity (hydrotropism; Eapen et al. 2005).

Higher plants can also detect the presence of other plants or objects in their neighborhood and respond accordingly (Baldwin 2010). For this purpose, they use volatile substances derived from terpenoids, fatty acid catabolites, aromatics and amino acids, which are released into the air. Plants that lose the ability to detect ethylene also lack the ability to sense the location of other plants and object in their surroundings. In addition, plants show tactile responses (thigmo- or haptotropism) when they touch an object. This is of an advantage for climbing plants as it facilitates the search for a support. The receptors are specific pits in the outer cell wall of the epidermis and they can even distinguish between different materials which e.g. prevents a vine to entwine a water jet (Isnard and Silk 2009).

All organs of sessile plants detect the gravity vector of the Earth but respond differently. Main stems generally grow upward (negative gravitropism; Yamamoto et al. 2002) while principal roots grow downward (positive gravitropism; Konings

Fig. 1.2 Positive and negative gravitropism of primary and secondary roots and shoots. Modified after Lüttge et al. (1994)

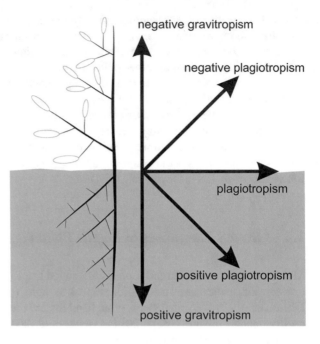

1995). Since these responses are so obvious they have been studied in depth for a long time (Pfeffer 1881; Bünning 1955; see Chap. 6). Things become more complicated when we look at lateral branches and roots as well as leaves (Fig. 1.2). Growth perpendicular to the gravity vector is called plagiotropism. Branches of plants orient themselves at an angle pointing upward (positive plagiotropism) and secondary roots pointing downwards (negative plagiotropism; Leakey 1990). One explanation for these phenomena is that two antagonistic reactions control the direction of growth, one being gravitropism and the other epinasty (Kang 1979; Edelmann et al. 2002). In branches and leaves epinasty is the downward bending resulting from a faster growth of the upper side than that of the lower side.

1.5   How Do Organisms Detect and Respond to Gravity?

In order to respond to the gravitational signal an organism needs a receptor capable of detecting the gravitational force. Two principles have evolved in the biota. In some organisms ranging from ciliates to algae to higher plants heavy cellular particles sediment and initiate a signal transduction chain (Limbach et al. 2005; Strohm et al. 2012). In the Chlorophyte *Chara*, heavy $BaSO_4$ crystals function as statoliths, both in the rhizoids, where they are responsible for positive gravitropism, and in protonemata, where they coordinate negative gravitropic upward growth (Sievers et al. 1996; cf. Chap. 4). Interestingly, $BaSO_4$ and $SrSO_4$ crystals are also used in the statocyst-like organelles of the ciliates *Loxodes* and *Remanella* (family

Loxodidae). In the roots and shoots of higher plants, amyloplasts are responsible as sedimenting statoliths (Blancaflor and Masson 2003; cf. Chap. 6). The molecular nature of the receptor, which senses the force of the accelerating statoliths, has not yet been revealed.

In cases where no heavy statoliths have been detected, the whole cytoplasmic content of a cell is thought to exert pressure onto the lower cell membrane where it could activate mechanosensitive channels which gate an influx of Ca^{2+} ions when stimulated (Häder and Hemmersbach 2017; cf. Chap. 3), a principle obviously used in some free-swimming ciliates. Here, the sedimenting cell content activates polarily distributed mechano-(gravi-)sensitive ion channels in the cell membrane, thereby, initiating signaling pathways which finally result in gravitactic orientation (Hemmersbach et al. 1996).

Due to the small size of the statolith and the low difference in the density of the cell content and the outer medium the force exerted on the gravireceptor is minute and close to the physical limit for detection given by Brownian movement. E.g. in the algae *Euglena* the cytoplasm of the cell has a specific density of about 1.045 g/mL. Using the cell volume V allows calculating the force F from the gravitational acceleration on Earth and the specific density difference.

$$F = V g \, \delta\rho$$

For this flagellate this yields a force of 0.49–1.23 pN. For *Arabidopsis* root cells a force of 0.017 pN has been calculated, for the *Chara* rhizoid barium sulfate vesicle 0.018 pN and for the ciliate *Bursaria* 11.8 pN, while the 100 times larger ciliate *Paramecium caudatum* produces a force of 128 pN (Häder et al. 2005).

In order to terminate in a physiological, biochemical or behavioral response, the signal has to be relayed from the gravireceptor to the effector in a sensory transduction chain (Häder and Hemmersbach 1997). This can involve the activation of proteins, enzymes or genes (Häder et al. 2017). In the case of motile microorganisms, this chain of events results in the reorientation of cilia or flagella which is instrumental in a gravitactic reorientation (Hemmersbach and Häder 1999). In higher plants, the situation is even more complicated since the receptor and the effector are located in different cells which are located at a considerable distance. While the gravireceptors are organized in the root tip columella, the growth response in the form of differential growth of the opposite root flanks (gravitropic bending) is elicited in the elongation zone well above the root tip. The messenger has been identified as the plant growth hormone auxin transported by specific PIN proteins (Friml et al. 2002). Depending on the complexity of the response and the distances over which the signal has to be relayed, the signal transduction chains can be of different lengths.

As in other stimulus signaling pathways regarding environmental stimuli, the gravity-related one consists of a number of individual steps: perception, transduction with amplification and response (Fig. 1.3).

The following chapters provide examples demonstrating common principles and differences in graviperception and gravisignaling pathways between evolutionary diverse organisms, ranging from unicellular plant and animal systems to

Fig. 1.3 Sensory transduction chain for graviperception and response involving a heavy sedimenting mass, a gravireceptor, signal amplification and finally a visible response

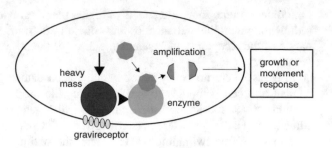

multicellular plant systems. Like in all other issues of this Springer Briefs series, the contribution of experimentation in the almost stimulus-free environment of space (microgravity) and on different microgravity-simulation platforms to our current knowledge are highlighted and potentials for future research and applications as well as open questions are addressed.

References

Baldwin IT (2010) Plant volatiles. Curr Biol 20:R392–R397

Bean B (1984) Microbial geotaxis. In: Colombetti G, Lenci F (eds) Membranes and sensory transduction. Plenum Press, London, pp 163–198

Bechert J (2009) Vergleichende Untersuchungen zur Gravitaxis und Phototaxis bei Ciliaten. PhD, Bonn

Blancaflor EB, Masson PH (2003) Plant gravitropism. Unraveling the ups and downs of a complex process. Plant Physiol 133:1677–1690

Bodanis D (2005) E=mc^2: a biography of the world's most famous equation. Bloomsbury Publishing, London

Briggs WR (2014) Phototropism: some history, some puzzles, and a look ahead. Plant Physiol 164:13–23

Buder J (1920) Neue phototropische Fundamentalversuche. Berichte der Deutschen Botanischen Gesellschaft 38:10–19

Bünning E (1955) Bewegungen. Fortschritte der Botanik 18:347–364

Buttinoni I, Volpe G, Kümmel F, Volpe G, Bechinger C (2012) Active Brownian motion tunable by light. J Phys Condens Matter 24:284129

Clegg MR, Maberly SC, Jones RI (2003) Chemosensory behavioural response of freshwater phytoplanktonic flagellates. Plant Cell Environ 27:123–135

Dusenbery DB (1985) Using a microcomputer and videocamera to simultaneously track 25 animals. Comput Biol Med 15:169–175

Eapen D, Barroso ML, Ponce G, Campos ME, Cassab GI (2005) Hydrotropism: root growth responses to water. Trends Plant Sci 10:44–50

Edelmann HG, Gudi G, Kühnemann F (2002) The gravitropic setpoint angle of dark-grown rye seedlings and the role of ethylene. J Exp Bot 53:1627–2634

Everett M, Thimann KV (1968) Second positive phototropism in the *Avena* coleoptile. Plant Physiol 43:1786–1792

Fenchel T, Finlay BJ (1984) Geotaxis in the ciliated protozoon *Loxodes*. J Exp Biol 110:17–33

Fiedler B, Börner T, Wilde A (2005) Phototaxis in the cyanobacterium *Synechocystis* sp. PCC 6803: role of different photoreceptors. Photochem Photobiol 81:1481–1488

Finlay BJ, Fenchel T (1986) Photosensivity in the ciliated protozoon *Loxodes*: pigment granules, absorption and action spectra, blue light perception, and ecological significance. J Protozool 33:534–542

Finlay B, Fenchel T, Gardener S (1986) Oxygen perception and O_2 toxicity in the freshwater ciliated protozoon *Loxodes*. J Eukaryot Microbiol 33:157–165

Fontana DR, Poff KL (1984) Effect of stimulus strength and adaptation on the thermotactic response of *Dictyostelium discoideum* pseudoplasmodia. Exp Cell Res 150:250–257

Fraenkel GS, Gunn DL (1961) The orientation of animals (Kineses, taxes and compass reactions). Dover Publication Inc., New York

Friml J, Wiśniewska J, Benková E, Mendgen K, Palme K (2002) Lateral relocation of auxin efflux regulator PIN3 mediates tropism in *Arabidopsis*. Nature 415:806–809

Häder D-P (1991) Strategy of orientation in flagellates. In: Riklis E (ed) Photobiology. The science and its applications. Plenum Press, New York, pp 497–510

Häder D-P, Hemmersbach R (1997) Graviperception and graviorientation in flagellates. Planta 203:7–10

Häder D-P, Hemmersbach R (2017) Gravitaxis in *Euglena*. In: Schwartzbach S, Shigeoka S (eds) *Euglena*: biochemistry, cell and molecular biology. Springer, Cham, pp 237–266

Häder D-P, Iseki M (2017) Photomovement in *Euglena*. In: Schwartzbach S, Shigeoka S (eds) *Euglena*: biochemistry, cell and molecular biology. Springer, Cham, pp 207–235

Häder D-P, Rosum A, Schäfer J, Hemmersbach R (1995) Gravitaxis in the flagellate *Euglena gracilis* is controlled by an active gravireceptor. J Plant Physiol 146:474–480

Häder D-P, Hemmersbach R, Lebert M (2005) Gravity and the behavior of unicellular organisms. Cambridge Univ. Press, Cambridge

Häder D-P, Braun M, Grimm D, Hemmersbach R (2017) Gravireceptors in eukaryotes – a comparison of case studies on the cellular level. npj Microgravity 3:13

Hamant O (2013) Widespread mechanosensing controls the structure behind the architecture in plants. Curr Opin Plant Biol 16:654–660

Hawking SW (2006) The theory of everything: the origin and fate of the Universe. Phoenix Books, Special Anniv

Hemmersbach R, Häder D-P (1999) Graviresponses of certain ciliates and flagellates. FASEB J 13: S69–S75

Hemmersbach R, Voormanns R, Briegleb W, Rieder N, Häder D-P (1996) Influence of accelerations on the spatial orientation of *Loxodes* and *Paramecium*. J Biotechnol 47:271–278

Hemmersbach-Krause R, Häder D-P (1990) Negative gravitaxis (geotaxis) of *Paramecium* – demonstrated by image analysis. Appl Micrograv Technol 4:221–223

Hemmersbach-Krause R, Briegleb W, Häder D-P (1991a) Dependence of gravitaxis in *Paramecium* on oxygen. Eur J Protistol 27:278–282

Hemmersbach-Krause R, Briegleb W, Häder D-P, Plattner H (1991b) Gravity effects on *Paramecium* cells: an analysis of a possible sensory function of trichocysts and of simulated weightlessness of trichocyst exocytosis. Eur J Protistol 27:85–92

Isnard S, Silk WK (2009) Moving with climbing plants from Charles Darwin's time into the 21st century. Am J Bot 96:1205–1221

Kamykowski D, Zentara SJ (1977) The diurnal vertical migration of motile phytoplankton through temperature gradients. Limnol Oceanogr 22:148–151

Kang BG (1979) Epinasty. In: Haupt W, Feinleib ME (eds) Physiology of movements. Encyclopedia of Plant Physiology. N.S. Springer-Verlag, Berlin, pp 647–667

Kim D (2013) Control of *Tetrahymena pyriformis* as a microrobot. PhD thesis, Drexel University

Kiss JZ, Mullen JL, Correll MJ, Hangarter RP (2003) Phytochromes A and B mediate red-light-induced positive phototropism in roots. Plant Physiol 131:1411–1417

Kohn F, Hauslage J, Hanke W (2017) Membrane fluidity changes, a basic mechanism of interaction of gravity with cells? Microgravity Sci Technol 29:337–342

Konings H (1995) Gravitropism of roots: an evaluation of progress during the last three decades. Acta Botanica Neerlandica 44:195–223

Kushner DJ (1985) The Halobacteriaceae. In: Woese CR, Wolfe RS (eds) Archaebacteria bacteria: a treatise on structure and function. Acad. Press, Orlando, pp 171–214

Leakey RRB (1990) *Nauclea diderrichii*: rooting of stem cuttings, clonal variation in shoot dominance, and branch plagiotropism. Trees-Struct Funct 4:164–169

Li G, Tam L-K, Tang JX (2008) Amplified effect of Brownian motion in bacterial near-surface swimming. Proc Natl Acad Sci 105:18355–18359

Limbach C, Hauslage J, Schäfer C, Braun M (2005) How to activate a plant gravireceptor. Early mechanisms of gravity sensing studied in characean rhizoids during parabolic flights. Plant Physiol 139:1030–1040

Lin W, Paterson GA, Zhu Q, Wang Y, Kopylova E, Li Y, Knight R, Bazylinski DA, Zhu R, Kirschvink JL (2017) Origin of microbial biomineralization and magnetotaxis during the Archean. Proc Natl Acad Sci:201614654

Liscum E, Askinosie SK, Leuchtman DL, Morrow J, Willenburg KT, Coats DR (2014) Phototropism: growing towards an understanding of plant movement. Plant Cell 26:38–55

Lohmann KJ, Lohmann CMF, Ehrhart LM, Bagley DA, Swing T (2004) Geomagnetic map used in sea-turtle navigation. These migratory animals have their own equivalent of a global positioning system. Nature 428:909–910

Lüttge U, Kluge M, Bauer G (1994) Botanik. VCH, Weinheim

Machemer H (1996) A theory of gravikinesis in *Paramecium*. Adv Space Res 17:11–20

Machemer H, Machemer-Röhnisch S, Bräucker R, Takahashi K (1991) Gravikinesis in *Paramecium*: theory and isolation of a physiological response to the natural gravity vector. J Comp Physiol A 168:1–12

Maree AFM, Panfilov AV, Hogeweg P (1999) Migration and thermotaxis of *Dictyostelium discoideum* slugs, a model study. J Theor Biol 199:297–309

Margenau H, Kestner N (1969) Theory of intermolecular forces. Pergamon Press, Oxford

Nultsch W (1975) Phototaxis and photokinesis. In: Carlile MJ (ed) Primitive sensory and communication systems. Academic Press, New York, pp 29–90

Ooya M, Mogami Y, Izumi-Kurotani A, Baba SA (1992) Gravity-induced changes in propulsion of *Paramecium caudatum*: a possible role of gravireception in protozoan behaviour. J Exp Biol 163:153–167

Pfeffer W (1881) Pflanzenphysiologie. Verlag Wilhelm Engelmann, Leipzig

Poff KL, Skokut M (1977) Thermotaxis by pseudoplasmodia of *Dictyostelium discoideum*. Proc Natl Acad Sci USA 74:2007–2010

Porterfield DM (1997) Orientation of motile unicellular algae to oxygen: oxytaxis in *Euglena*. Biol Bull 193:229–230

Ruppel NJ, Hangarter RP, Kiss JZ (2001) Red-light-induced positive phototropism in *Arabidopsis* roots. Planta 212:424–430

Sato A, Sasaki S, Matsuzaki J, Yamamoto KT (2015) Negative phototropism is seen in *Arabidopsis* inflorescences when auxin signaling is reduced to a minimal level by an Aux/IAA dominant mutation, axr2. Plant Signal Behav 10:e990838

Sievers A, Buchen B, Hodick D (1996) Gravity sensing in tip-growing cells. Trends Plant Sci 1:273–279

Simmons SL, Sievert SM, Frankel RB, Bazylinski DA, Edwards KJ (2004) Spatiotemporal distribution of marine magnetotactic bacteria in a seasonally stratified coastal salt pond. Appl Environ Microbiol 70:6230–6239

Steidinger KA, Tangen K (1996) Dinoflagellates. In: Tomas CR, Hasle GR, Syvertsen EE (eds) Identifying Marine Diatoms and Dinoflagellates. Academic Press Inc., London, pp 387–585

Streeter J (2016) God and the history of the Universe. Wipf and Stock Publishers, Eugene

Strohm AK, Baldwin KL, Masson PH (2012) Molecular mechanisms of root gravity sensing and signal transduction. Wiley Interdiscip Rev Dev Biol 1:276–285

Todd P (2007) Gravity-dependent phenomena at the scale of the single cell. Gravit Space Res 2:95–113

Votta JJ, Jahn TL (1972) Galvanotaxis of *Chilomonas paramecium* and *Trachelomonas volvocina*. J Protozool 19(Suppl):43

Wadhams GH, Armitage JP (2004) Making sense of it all: bacterial chemotaxis. Nat Rev Mol Cell Biol 5:1024–1037

Whitaker BD, Poff KL (1980) Thermal adaptation of thermosensing and negative thermotaxis in *Dictyostelium*. Exp Cell Res 128:87–93

Yamamoto H, Yoshida M, Okuyama T (2002) Growth stress controls negative gravitropism in woody plant stems. Planta 216:280–292

Zhenan M, Shouyu R (1983) The effect of red light on photokinesis of *Euglena gracilis*. In: Tseng CK (ed) Proceedings of the Joint China-U.S. Phycology Symposium. Sci. Press, Beijing, pp 311–321

Chapter 2
Methods for Gravitational Biology Research

Ruth Hemmersbach, Donat-Peter Häder, and Markus Braun

Abstract To study the impact of gravity on living systems on the cellular up to the organismic level, a variety of experimental platforms are available for gravitational biology and biomedical research providing either an almost stimulus-free microgravity environment (near weightlessness) of different duration and boundary conditions. The spectrum of real-microgravity research platforms is complemented by devices which are used to either increase the gravity level (centrifuges) or modify the impact of gravity on biological systems (clinostats and random-positioning machines)—the so-called ground-based facilities. Rotating biological samples horizontally or in a two- or three-dimensional mode is often used to randomize the effect of gravity in the attempt to eliminate the gravity effect on sensing mechanisms and gravity-related responses. Sophisticated centrifuges have been designed allowing studies from cells up to humans, either on ground under hypergravity conditions (> 1 g) or in space, where they offer the chance to stepwise increase the acceleration force from 0 g (microgravity) to 1 g or higher and *vice versa*. In such a way, centrifuges are used to determine threshold values of gravisensitivity and to unravel molecular and cellular mechanisms of gravity sensing and gravity-related responses. By using the whole spectrum of experimental platforms, gravitational biologists gain deep insight into gravity-related biological processes and continuously increase our knowledge of how gravity affects life on Earth.

Keywords Clinostat · Random Positioning Machine · Parabolic flight · ISS · Space shuttle · Satellite

2.1 Introduction

Gravity is a unique environmental stimulus, constantly acting, thus, having shaped life during evolution. Consequently, the question arises about its impact, how it affects fundamental physiological processes. Most organisms have developed a specific gravisensor system and use gravity for orientation, but gravity also generally affects physiological, cellular and molecular processes, both best investigated in the

M. Braun et al., *Gravitational Biology I*, SpringerBriefs in Space Life Sciences, https://doi.org/10.1007/978-3-319-93894-3_2

stimulus-free environment of space in the absence of gravity-induced phenomena such as sedimentation, buoyancy and convection.

Studies in microgravity opened new scientific perspectives and a new field of experimentation. We are now able to bring an organism in a very new and unique environment that it has not yet experienced before. What does this mean for the organism? Does it cause stress when you take away an environmental cue that was used for orientation? Does an organism experience less stress when a structure which has been under tension is now relaxing or which might have sedimented on membranes will now float in the absence of gravity? What does a three-dimensional free-floating environment mean for a cell which has before been attached to a layer? What is the effect of mechanical unloading with respect to function and differentiation? How quickly will a biological system respond, will it adapt, what is the minimum amount of gravitational acceleration necessary to initiate a gravity-related response? Are there sensitive windows in development during which an organism is more sensitive to changes of environmental parameters like gravity? In order to address these fundamental questions which also bear—as we will see later—application potentials, dedicated platforms for such research are needed.

Due to the presence of masses in our universe, it is impossible to achieve zero gravity (real weightlessness). Even on the ISS circling Earth at an altitude of about 350 km, the level of gravity is only 8% less than on the Earth's surface. It is the velocity of the space station that creates a centrifugal force which exactly compensates the centrifugal force of the gravitational pull of the Earth that results in a free-fall situation which we call microgravity (near weightlessness) due to some residual acceleration forces in a range of 10^{-2}–10^{-6} g.

Limited access to space flight and high costs motivated developments to achieve—to at least to some extent—microgravity conditions on ground. Today, ground-based studies have great importance in gravitational space biology and human physiology research. They increasingly contribute to our understanding of how biological systems (from cells to humans) sense gravity and to study the consequences when the influence of this fundamental force is lacking, to study the impact on health and signaling cascades, but also to study adaption mechanisms to this new environmental condition. In this chapter, we give an overview of available ground-based microgravity simulators and platforms providing increased gravity levels, rounded up by a comprehensive summary of platforms enabling unique experimentation in real microgravity. We will focus on the underlying principles, boundary conditions and the experimental possibilities.

2.2 Microgravity Simulators—Efforts to Mimic the Effects of Weightlessness

2D, 3D, fast and slow rotating clinostats, rotating wall vessel, magnetic levitator, Random Positioning Machine—if you are looking for devices in order to mimic the effects of weightlessness on ground you will find a catalogue of possibilities. How do they differ, what is the method of choice, what is the best suitable facility for my scientific object and question?

The idea to alter the influence of gravity and study the impact on basic biological mechanisms is quite old and started experimentally at the end of the eighteenth century with plants due to their easy observable gravitropic responses. By putting them horizontally the restoration of their original growth direction by gravitropic responses becomes obvious—roots growing downwards and shoots growing upwards. If such kind of arrangement is equipped with a motor and the plant is rotated around an axis perpendicular to the gravity vector, the unidirectional influence of gravity is turned into an omnilateral stimulation which in many cases abolishes the gravitropic response. Under optimal conditions the gravitropic stimulus is neutralized, like in real microgravity, a situation called simulated weightlessness (microgravity). In both cases a plant will no longer show gravitropism, but what has happened with respect to the underlying gravisensory mechanism—is it permanently stimulated or does it receive no further input? We will come back to this point later. Such kind of experimental arrangement is called a clinostat.

A 2D clinostat has only a single rotation axis. Wolfgang Briegleb used the 2D clinostat principle for studying the effect of weightlessness on small plants and animals and cells. He postulated that speeding up and thus transforming a slow rotating clinostat, normally rotated with 1–2 rpm (revolutions per minute), into a fast-rotating one (in the range of 60–90 rpm) will optimize the simulation of weightlessness conditions (Briegleb 1988; Klaus et al. 1998). Furthermore, not only speed but also the effective radius (diameter) has to be considered. Under optimal conditions, the diameter of the sample containers is kept small (in the range of a few mm) and the objects are placed in the center of rotation in order to keep residual accelerations as minimal as possible. The latter concerns thresholds for gravity stimulus perception of the respective organism, which are in most cases not known. A 2D clinostat constantly runs in one direction inducing a static change of the gravity vector in relation to the sample. Sedimentation is thereby prevented and small bodies (e.g. single cells or statoliths within cells) describe floating circles in the media comparable to the floating conditions in real microgravity. The speed of rotation determines the circles' diameter; the faster the rotation, the smaller the circles; too fast rotation, however, results in radial accelerations. Let us transfer this idea to statoliths in roots or rhizoids and imagine their movements depending on the speed of rotations. Having done so, Hensel and Sievers (1980) demonstrated by morphological studies of slowly clinorotated roots (1–2 rpm) strong damages of the statocytes on the ultrastructural level, e.g. revealed by a considerable increase of the lytic compartment. They related these changes to the continuously changing direction of the gravity vector, which is different to the situation in real microgravity.

2D clinostats have been adapted to several experimental demands (for review see Brungs et al. 2016): clinostats for suspended or adherent organisms and cell cultures, for aquatic systems and in combination with online analyses using photomultipliers or microscopy (Fig. 2.1).

Assuming that two rotation axes provide more complex ways to average the influence of the gravity vector and simulate weightlessness more perfectly, 3D

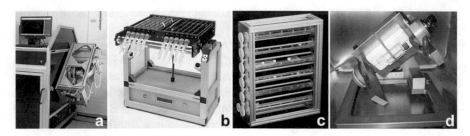

Fig. 2.1 Examples of ground-based facilities to simulate microgravity conditions: Various 2D clinostats based on the principle of fast and continuous clinorotation around one axis of rotation: (**a**) live-cell imaging fluorescence clinostat microscope, (**b**) pipette-based clinostat for the exposure of cell suspensions, (**c**) slide-flask clinostat for adherent cell cultures. Random-positioning machine mostly used in a random speed and random direction operational mode, resulting in a disorientation of the exposed samples (**d**)

clinostats and the Random Positioning Machines (RPM) have come into use. They are characterized and operated with two independently rotating frames mounted in a gimbal manner (Hoson et al. 1996; van Loon 2007). An algorithm controls the motors with respect to acceleration or directional changes. Commonly, 3D clinostats are continuously rotating but changing the velocity at random. In a RPM, not only the velocity but in addition also the direction of rotation is randomized.

Comparative studies—also in real microgravity—are necessary to understand the differences and to validate the quality of the simulation (Herranz et al. 2013). Here, we will give some examples to demonstrate this kind of approach.

To critically assess the assumption whether a second rotation axis and sophisticated modes of operation provide a more perfect simulation, Krause et al. (2018) studied the dynamics of the actin-dependent movements of statoliths in the rhizoids of *Chara* (cf. Chap. 4). The role of gravity in this process was already investigated in real microgravity in a MAXUS sounding rocket mission; thus, data for verification and validation were available and could be compared to data from 2D and 3D clinorotation. Fast rotational speeds in the range of 60–85 rpm in 2D and 3D modes resulted in a similar kinetics of statolith displacement as compared to real µg, while slower clinorotation (2–11 rpm) caused a reduced one. The addition of a second rotation axis clearly did not increase the quality of microgravity simulation, however, increased non-gravitational effects such as an increase in the level of vibration (with multiple potential side effects). Thus, for *Chara* rhizoids, fast 2D clinorotation is the most appropriate microgravity simulation method for investigating its graviperception mechanism.

Hauslage et al. (2017) visualized shear and hydrodynamic forces in various ground-based facilities by using dinoflagellates as bioassay and mechanosensitive reporter systems. *Pyrocystis noctiluca* populations were exposed on a Random Positioning Machine either operating as 2D clinostat (constant rotation around one axis with 60 rpm) or in a random positioning mode (two axes with random velocity and direction). Shear stress due to hydrodynamic forces leads to a deformation of the cell membrane, induces intracellular signaling triggering an increase in cytosolic Ca^{2+} and in turn the reaction between luciferin and luciferase resulting in the emission of light. Thus, the signal intensity provides valuable information about the shear stress induced by the different microgravity simulation methods. The data show that exposure on the RPM resulted in a higher mechanical stress for the dinoflagellates than during constant clinorotation (Fig. 2.2). This proved 2D clinorotation as a low shear stress environment.

Rotating wall vessels (RWVs) or rotating bioreactors are further methods which are frequently being used to neutralize sedimentation in aquatic systems. Although these methods have been widely used for studying cell cultures, protists and other small aquatic organisms, its ability to mimic weightlessness still has to be demonstrated by comparative studies in real microgravity (Schwarz et al. 1992).

Magnetic levitation is offered as a further approach to achieve microgravity in a ground laboratory. However, in the case of biology, the effects of the magnetic

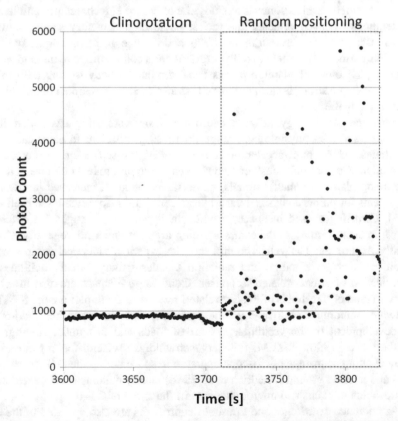

Fig. 2.2 The capacity of bioluminescence as a result of a mechanical stimulation of dinoflagellates was used as bioassay. Exposure of *Pyrocystis noctiluca* on a rotating device, either operated as 2D clinostat (constantly running around one axis) or as random positioning machine (rotating around two axes at a random speed and random direction mode) revealed a differential stress response indicated by the number of photons emitted, modified after Hauslage et al. (2017)

field itself have to be considered, which had a strong impact on the alignment of gravitactic organisms such a *Euglena* and *Paramecium* (Hemmersbach et al. 2014) and mask the expected behavior known from experiments in real microgravity (Häder et al. 1990; Hemmersbach-Krause et al. 1993) (cf. Chap. 3).

2.3   Centrifuges—The Benefit of Hypergravity in Gravitational Biology Research

Centrifugation is an experimental approach to artificially and experimentally increase the influence of gravitational acceleration greater than the one normally acting on the surface of Earth (1 g). Hypergravity is a tool to obtain new insights into gravity-related molecular and physiological mechanisms. Centrifugation experiments are widely used as controls to study the effects of launch, reentry and landing accelerations of space vehicles like rocket payloads and reentry satellites, which allows discriminating between these effects and the one of the microgravity conditions. Furthermore, the potential of hypergravity as a countermeasure method against negative physiological adaption processes of the human body to long-term microgravity, such as a strong bone and muscle loss and decrease in cardiac functioning, is under investigation.

In gravitational biology, different centrifuge designs are being used, operating at low speed and thus physiological range up to 20 g, different to usual laboratory centrifuges, where high accelerations are applied for separation during sample preparation. Custom-built desktop centrifuges are appropriate to culture cells and developing plants or (small) animals, while larger samples, increased hardware or instruments for online analyses demand large centrifuges (for review see Frett et al. 2016). Centrifuges used in space provide an appropriate in-flight 1-g reference control in close vicinity to the samples which are kept under microgravity. Consequently, experimental as well as control samples experience identical environmental conditions such as vibration and radiation besides gravity, which facilitates the identification of gravity-related responses. Centrifuges in space are also in use for determination of thresholds of gravity-related processes. Examples are the STATEX hardware including a centrifuge for the performance of an experiment addressing the development of the vestibular system of toads and fish under microgravity (Neubert et al. 1996). NIZEMI (Niedergeschwindigkeits-Zentrifugen-Mikroskop), a slow rotating centrifuge microscope operated in space during the IML2 mission in 1994 and later on ground (Friedrich et al. 1996) clearly demonstrated the existence of thresholds in plants and microorganisms in the range of 0.1–0.3 g.

1-g reference centrifuges and threshold centrifuges are also operated in the drop tower or parabolic flights of airplanes and rockets (TEXUS, MAXUS) as well as in the ESA's European Modular Cultivation System EMCS and ESA's BIOLAB, both facilities aboard the ISS and designed to carry out experiments in biomedical research (Brinckmann 2005). The fact that thresholds for gravity-related responses exist, as revealed by centrifuge experiments in space, indicate physiological rather than a pure passive mechanism of a response which was e.g. a long-lasting debate in case of gravitaxis of protists (unicellular organisms) (see Chap. 3). Furthermore, it shows that we cannot just simply extrapolate data obtained in hypergravity to "0 g" and predict the result. Large centrifuges such as the Short Arm Human Centrifuge at DLR meet the demands of life science researchers for complex training and bio-medical examinations under hypergravity conditions. Such a large centrifuge system provides a great platform for biomedical and neurophysiological research, but is also used by cell biologists, who started using it to operate various instruments like a life cell imaging microscope under hypergravity conditions.

With respect to exploration and the need of closed biological life support systems, plants and (lower) animals might become essential parts of bioregenerative life support systems in order to provide nutrients and energy on long-term missions or on other planets. It is therefore of great importance to investigate the effects of lunar (ca. 0.16 g) or Martian (ca. 0.38 g) gravity on plant/animal metabolism, growth, proliferation and development as well as on human beings. Centrifugation in space provides these conditions. EU:CROPIS, a compact satellite scheduled to fly in 2018, will provide lunar and Martian gravity conditions for 6 months each to study the impact of these gravity levels on the performance of a biological life support system, further equipped with a special trickling filter unit for urine degradation, a food production unit and a *Euglena* compartment for oxygen pro-duction (cf. Chap. 8).

2.4   From Drop Tower to ISS—Biology in Free Fall

Several excellent experimental platforms offer real microgravity conditions for gravitational biology research: drop towers, parabolic flights, sounding rocket flights, satellites and the International Space Station ISS (Fig. 2.3). They differ in the time of microgravity provided and the quality of microgravity that can be achieved. Thanks to the excellent cooperation between the scientists, technicians and astronauts highly sophisticated research hardware has been developed and operated in space. After several decades of space biology research, nowadays, a huge amount of experience is available for performing biological experiments on cells, tissues, and organisms including human beings. Cultivation and fixation, life support systems as well as online microscopic and kinetic studies are daily work for the astronauts on the ISS. This chapter briefly describes the platforms that provide real microgravity conditions with special focus on their characteristics and boundary conditions. Some of them even provide opportunities for student experiments such as the "Drop, Fly, and Spin your Thesis" or the REXUS/BEXUS student rocket and balloon program at Esrange near Kiruna in Northern Sweden, jointly organized by the Germany DLR Space Administration and the Swedish National Space Board.

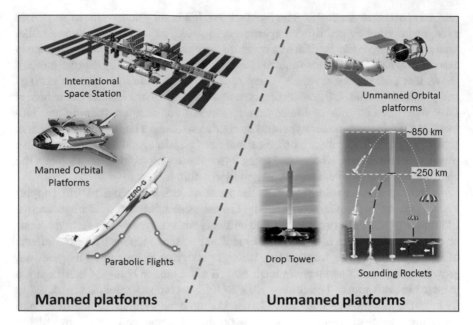

Fig. 2.3 Manned and unmanned Platforms providing real microgravity conditions

2.4.1 Drop Tower

A high quality of microgravity of about 10^{-6} g is achieved during the 110 m free fall in the vacuum tube of the drop tower at the Zentrum für Angewandte Raumfahrttechnologie und Mikrogravitation (ZARM) in Bremen, Germany. Favorable features are the daily accessibility, multiple launches and drops per week, thereby, guaranteeing a sufficient amount of replicates and a flexible experiment adaptation. Microgravity period of 4.74 s or even 9.3 s, the latter by using a catapult system, provide ample time for a surprisingly great number of fast cell biological and physiological processes (Könemann et al. 2015). Although an experiment time of 4.74 s is rather short, the fact that no hypergravity phase is involved is of great advantage especially for sensitive and quickly responding processes.

Using the catapult, however, unavoidably comes with an initial phase of high accelerations of up to 30 g for 0.26 s, to propel the drop capsule up into the evacuated tower shaft. These boundary conditions have to be taken into account and their effects on the biological sample have to be carefully assessed by control experiments. Biologists are frequently using the drop tower e.g. to study fast molecular, cellular and physiological responses of cells and organisms after the offset of gravity.

2.4.2   Parabolic Plane Flights

Microgravity times of several minutes are provided by very special parabolic flight maneuvers with airplanes. The term parabolic flight describes a flight maneuver that enables an aircraft, rocket or spacecraft to follow a free-fall ballistic Keplerian trajectory (Ruyters and Friedrich 2006b; Pletser et al. 2015). Parabolic flights of aircrafts have become a working horse for training astronauts, hardware testing and preparation for ISS experiments as well as for stand-alone biomedical and psychological experiments on human subjects and a broad variety of biological experiments.

Since 2015, the Airbus A310 ZERO-G, which replaced the old A300 Zero-G aircraft, is the largest aircraft for European microgravity research (Pletser et al. 2015). It is operated by Novespace, a subsidiary of the French National Space Center (CNES), with the European Space Agency ESA and DLR, the German Space Administration, as frequent customers. Usually, a parabolic flight campaign consists of 31 parabolas flown at each of the three consecutive flight days. In total, 93 microgravity (µg) phases of approximately 22 s add up to 10 min of µg with a residual acceleration of about 10^{-2} g. The parabolic flight maneuvers can be flown in such a way that partial g-levels in the range of lunar (0.16 g) and Martian (0.38 g) gravity are achieved for approx. 22 s (Pletser et al. 2012). Of great advantage for scientists is the fact that the investigator can bring basically his own familiar—even bigger—lab equipment on board and can perform the experiment himself during the flight, thus, being able to monitor and operate his experiment, change experimental parameters or the experimental set-up during the flight. A major disadvantage might be the flight profile consisting of alternating phases of hypergravity phases of up to 1.8 g, microgravity and 1 g acceleration in between the parabolic flight phases. A careful assessment of the results and proper control experiments are necessary to distinguish clearly between microgravity-induced effects, hypergravity effects and the effects of vibrations. Nevertheless, these microgravity periods are sufficient to address numerous questions in the area of gravitational biology ranging from the impact of microgravity on the cellular level, impacts on physiological parameters and the behavior of organisms up to biomedical and neurobiological studies on human subjects in microgravity.

2.4.3   Sounding Rockets and Suborbital Platforms

Microgravity in the range of minutes is provided by parabolic flights of rockets like MASER, TEXUS and MAXUS (Ruyters and Friedrich 2006a; Seibert and Battrick 2006). Today, sounding rockets are frequently used in microgravity research all over the world. In Germany, the TEXUS Sounding Rocket Programme (Technologische EXperimente Unter Schwerelosigkeit) started 1977 and 56 TEXUS rockets have

been launched until 2018. In 1990, the first European sounding rocket MAXUS
was launched, followed by the first MiniTEXUS rocket in 1993. With Mapheus
(Materialphysikalische Experimente unter Schwerelosigkeit), DLR started another
rocket program in 2009. The research rockets are launched from the European rocket
launch site ESRANGE near Kiruna in Northern Sweden. On its ballistic flight,
microgravity conditions in a range of 10^{-4} g last for about 3.5 min on a MiniTEXUS
flight, about 6 min on a TEXUS/Mapheus flight and about 13 min on a MAXUS
flight. The payload part of the rocket consisting of the Experiment Modules and the
Recovery and Service System decends on a parachute and samples are transported
back to the science labs at the launch site within 1–2 h by helicopter. Scientists can
directly monitor and control their experiments via telecommanding. Especially
appreciated by biologists is the late access allowing a loading of samples into the
payload only a few hours before launch and early retrieval of samples after landing.
Hence, a broad spectrum of basic biological and biomedical research was performed
with a variety of organisms, in most cases accompanied by centrifugation control
experiments to assess the influence of launch vibrations and accelerations in a range
of 6–12 g.

Alternatively, longer and more flat parabolas with a suborbital trajectory up to an
altitude of 100 km can be flown providing a continuous microgravity environment
for about 3 min. With such suborbital platforms, several US companies are aiming
to make microgravity experience available to space tourists. Such a commercial
platform might also add to the spectrum of microgravity research platforms;
however, the reliability and acceptability for space biology experiments still needs
to be demonstrated. Some considerations on this topic have been published e.g. by
Karmali and Shelhamer (2010).

2.4.4 Orbital Platforms—Space Shuttle, Satellites and the International Space Station

Long-term effects of microgravity in the range of weeks or months can only be
studied in low Earth orbit on board of satellites and human-tended space laboratories such as NASA's Space shuttles (which retired in 2011), Russian space
stations and the International Space Station ISS. In 1957 the dog Laika was the
first animal astronaut that surrounded Earth onboard a Sputnik-2 satellite. Laika did
not survive but two other dogs following in Sputnik-5 provided evidence that living
organisms can survive in low Earth orbit. Since then, numerous Russian Bion
satellites and American biosatellites housed a great variety of animals like snails,
worms, spiders, bees, frogs, fish, birds, mice and rats and all kinds of higher plants,
fungi, lichens, mosses, ferns and microorganisms. Even complex live support
systems have been developed and tested, which will increasingly become important

to complement physico-chemical life-support systems on future long-term space missions (cf. Chap. 9).

Satellites provided research platforms to study the effects of space conditions, mainly weightlessness and radiation, on living organisms. Encouraged by the first results, in 1961 Jurij Gagarin became the first human astronaut flying and surviving in low Earth orbit 300–400 km above the ground. Today, human beings have been working and living in space well protected from the harsh conditions of space—vacuum, extreme temperatures and high radiation exposure—in different kinds of return capsules, in Saljut and Kosmos space stations, Skylab, Space Shuttles, the Russian space station MIR and more recently in the Chinese space lab Tiangong. Microgravity research in space in the field of human physiology, neuroscience, animal physiology, plant biology, radiation biology, astrobiology, exobiology and microbiology always was and will be very much relying on international cooperation, mostly unimpressed by short-lived political issues.

The International Space Station is the largest international science and technology project ever undertaken representing the by far largest microgravity research platform that ever existed involving the United States, Russia, Japan, Canada and 10 member states of the European Space Agency including Germany, France and Italy. Based on the political decision for a symbol of international peaceful cooperation in the Low Earth orbit, all Space Station partners have invested greatly in this unique endeavor. Although the various space agencies may emphasize different goals and research objectives in the use of the ISS, they are all unified in the (1) recognition of the ISS as an education platform to encourage, inspire and ultimately motivate today's youth to pursue careers in math, science and engineering, (2) advancement of knowledge in all areas of human physiology, biology, material and physical sciences in a very unique radiation, microgravity and isolation environment and (3) translation of that knowledge to health, advanced product developments, socio-economic and environmental benefits to our lives on Earth.

For a general description and information on the ISS check the following DLR, ESA, NASA websites. A list of all research facilities onboard ISS is continuously updated at http://www.dlr.de/dlr/desktopdefault.aspx/tabid-10301/460_read-534/#/gallery/503, http://www.esa.int/Our_Activities/Human_Spaceflight/International_Space_Station, https://www.nasa.gov/mission_pages/station/main/index.html. The assembly of the ISS started in 1998 and was completed in 2010 providing pressurized modules developed by the NASA (USA), Roscosmos (Russia), JAXA (Japan) and ESA (Europe) and external platforms for science, technology demonstrations, education and a test bed for human space exploration beyond the low Earth orbit.

A great number of various types of specific racks offer experimental conditions and equipment for systematic studies in long-term microgravity. NASA provides utilization statistics and a history of research projects at the following website: http://www.nasa.gov/pdf/695701main_Current_ISS_Utilization_Statistics.pdf. Updates

on ISS activities, research and accomplishments can be found at: http://www.nasa.
gov/mission_pages/station/main/index.html. For more detailed information on
European participation and facts about the ISS please check: http://www.esa.int/
Our_Activities/Human_Spaceflight/International_Space_Station/About_the_Inter
national_Space_Station. The Erasmus Experiment Archive is ESA's database for
European funded or co-funded experiments not only on ISS but also on other
microgravity platforms and in microgravity ground-based facilities: http://eea.
spaceflight.esa.int/portal.

Due to the fact that spaceflight-related projects are costly and research opportunities are scarce, great efforts are undertaken to coordinate scientific utilization of the
ISS in a most efficient way by coordination through international and bilateral
working groups consisting of the Space Station partners and other leading space
agencies like DLR (Germany), CNES (France) and ASI (Italy).

Since the ISS is the only available platform of its kind with regard to humans
as subjects for health-related and fundamental biological research, the long-term
microgravity, isolation and radiation environment, sophisticated research facilities
with significant power and data resources, highly efficient and extensive utilization
and exploitation of this unique research platform are essential for the next decade—
and are mandatory for preparing human exploratory missions to Moon and Mars and
beyond.

2.5 Conclusions

In the last decades, our knowledge in the field of gravitational biology has made
considerable progress thanks to an increasing number of microgravity platforms
providing almost stimulus-free environments of different quality and duration.
Microgravity research opportunities, however, are still rare, costly and require a
complex organization, preparation and in most cases highly specific experiment
hardware for habitation, cultivation, fixation and sample analyses—well adapted to
the respective platform. Various microgravity simulation methods complement-
ing the real microgravity platforms have been invented for gravitational biology
research aiming to neutralize the effects of gravity on biological systems and alter
gravity conditions on ground. These methods are valuable tools for stand-alone
experiments, for proving new concepts and hypotheses, preparing microgravity
experiments, verifying microgravity results and testing hardware. However, thoroughly assessing all kinds of side effects and boundary conditions is required for
each biological sample. With the availability of space stations like the ISS and
future stations in low Earth orbit and beyond, the way has been paved for long-term
experimentation in microgravity yielding great opportunities for unraveling the
impact of gravity on life on Earth and preparing humans to explore the solar
system.

References

Briegleb W (1988) Ground-borne methods and results in gravitational cell biology. Physiologist 31: S44–S47

Brinckmann E (2005) ESA hardware for plant research on the International Space Station. Adv Space Res 36:1162–1166

Brungs S, Egli M, Wuest SL, Christianen PCM, van Loon JWA, Ngo Anh TJ, Hemmersbach R (2016) Facilities for simulation of microgravity in the ESA ground-based facility programme. Microgravity Sci Technol 28:191–203

Frett T, Petrat G, van Loon JJ, Hemmersbach R, Anken R (2016) Hypergravity facilities in the ESA ground-based facility program—current research activities and future tasks. Microgravity Sci Technol 28:205–214

Friedrich ULD, Joop O, Pütz C, Willich G (1996) The slow rotating centrifuge microscope NIZEMI—a versatile instrument for terrestrial hypergravity and space microgravity research in biology and materials science. J Biotechnol 47:225–238

Häder D-P, Vogel K, Schäfer J (1990) Responses of the photosynthetic flagellate, *Euglena gracilis*, to microgravity. Appl Micrograv Technol 3:110–116

Hauslage J, Cevik V, Hemmersbach R (2017) *Pyrocystis noctiluca* represents an excellent bioassay for shear forces induced in ground-based microgravity simulators (clinostat and random positioning machine). NPJ Microgravity 3:12

Hemmersbach R, Simon A, Waßer K, Hauslage J, Christianen PC, Albers PW, Lebert M, Richter P, Alt W, Anken R (2014) Impact of a high magnetic field on the orientation of gravitactic unicellular organisms—a critical consideration about the application of magnetic fields to mimic functional weightlessness. Astrobiology 14:205–215

Hemmersbach-Krause R, Briegleb W, Häder D-P, Vogel K, Grothe D, Meyer I (1993) Orientation of *Paramecium* under the conditions of weightlessness. J Eukaryot Microbiol 40:439–446

Hensel W, Sievers A (1980) Effects of prolonged omnilateral gravistimulation on the ultrastructure of statocytes and on the graviresponse of roots. Planta 150:338–346

Herranz R, Anken R, Boonstra J, Braun M, Christianen PC, de Geest M, Hauslage J, Hilbig R, Hill RJ, Lebert M (2013) Ground-based facilities for simulation of microgravity: organism-specific recommendations for their use, and recommended terminology. Astrobiology 13:1–17

Hoson T, Kamisaka S, Buchen B, Sievers A, Yamashita M, Masuda Y (1996) Possible use of a 3-D clinostat to analyze plant growth processes under microgravity conditions. Adv Space Res 17:47–53

Karmali F, Shelhamer M (2010) Neurovestibular considerations for sub-orbital space flight: a framework for future investigation. J Vestib Res 20:31–43

Klaus DM, Todd P, Schatz A (1998) Functional weightlessness during clinorotation of cell suspensions. Adv Space Res 21:1315–1318

Könemann T, Kaczmarczik U, Gierse A, Greif A, Lutz T, Mawn S, Siemer J, Eigenbrod C, von Kampen P, Lämmerzahl C (2015) Concept for a next-generation drop tower system. Adv Space Res 55:1728–1733

Krause L, Braun M, Hauslage J, Hemmersbach R (2018) Analysis of statoliths displacement in *Chara* rhizoids for validating the microgravity-simulation quality of clinorotation modes. Microgravity Sci Technol 30(3):229–236

Neubert J, Schatz A, Briegleb W, Bromeis B, Linke-Hommes A, Rahmann H, Slenzka K, Horn E (1996) Early development in aquatic vertebrates in near weightlessness during the D-2 mission STATEX project. Adv Space Res 17:275–279

Pletser V, Winter J, Duclos F, Bret-Dibat T, Friedrich U, Clervoy J-F, Gharib T, Gai F, Minster O, Sundblad P (2012) The first joint European partial-g parabolic flight campaign at moon and mars gravity levels for science and exploration. Microgravity Sci Technol 24:383–395

Pletser V, Rouquette S, Friedrich U, Clervoy J-F, Gharib T, Gai F, Mora C (2015) European parabolic flight campaigns with Airbus ZERO-G: looking back at the A300 and looking forward to the A310. Adv Space Res 56:1003–1013

Ruyters G, Friedrich U (2006a) From the Bremen drop tower to the international space station ISS—ways to weightlessness in the German space life sciences program. Signal Transduct 6:397–405

Ruyters G, Friedrich U (2006b) Gravitational biology within the German space program: goals, achievements, and perspectives. Protoplasma 229:95–100

Schwarz RP, Goodwin TJ, Wolf DA (1992) Cell culture for three-dimensional modeling in rotating-wall vessels: an application of simulated microgravity. J Tissue Cult Methods 14:51–57

Seibert G, Battrick BT (2006) The history of sounding rockets and their contribution to European space research. ESA Publications Division, Noordwijk

van Loon JJ (2007) Some history and use of the random positioning machine, RPM, in gravity related research. Adv Space Res 39:1161–1165

Chapter 3
Gravitaxis in Flagellates and Ciliates

Donat-Peter Häder and Ruth Hemmersbach

Abstract Motile microorganisms such as flagellates and ciliates use the gravity vector of the Earth to adjust their position in the water column. Oriented movement by gravity is called gravitaxis and can be positive (downward swimming) or negative (upward swimming). In addition, some ciliates modify their velocity according to the swimming direction (gravikinesis). Earth-bound research and experimentation under simulated and real microgravity have revealed that a heavy mass such as a statolith or the whole cell content presses onto a gravireceptor which perceives the signal. In some cases mechanosensitive ion channels have been identified as gravireceptors. The activation of the receptor results in a cascade of reactions which amplify the signal and result in a steering response changing the direction of movement.

Keywords Ciliates · Flagellates · Gravitaxis · Gravikinesis · Mechanosensitive ion channels · Sensory transduction chain

3.1 Introduction

Motile microorganisms such as flagellates and ciliates orient themselves with respect to a number of physical and chemical factors in their environment to optimize the conditions for growth and reproduction by using the light direction (phototaxis) for vertical migrations in the water column (Fraenkel and Gunn 1961; Häder 1991; Fiedler et al. 2005), the magnetic field of the Earth (magnetotaxis) (Simmons et al. 2004), thermal gradients (thermotaxis) (Maree et al. 1999) and chemical gradients of attractants and repellents (chemotaxis) (Clegg et al. 2003; Wadhams and Armitage 2004).

Graviorientation has been studied in many motile microorganisms such as flagellates *Tetraselmis, Dunaliella, Prorocentrum, Cryptomonas* and *Gymnodinium* (Rhiel et al. 1988; Eggersdorfer and Häder 1991; Richter et al. 2007) and ciliates such as *Paramecium, Tetrahymena, Bursaria, Stylonychia* and *Loxodes* (Finlay et al. 1993; Hemmersbach and Donath 1995; Hemmersbach and Bräucker 2002), The resulting vertical migrations produce pronounced phytoplankton distributions in

M. Braun et al., *Gravitational Biology I*, SpringerBriefs in Space Life Sciences, https://doi.org/10.1007/978-3-319-93894-3_3

the water column which may be affected by other external factors such as light and temperature and can be disturbed by the action of wind and waves (Piazena and Häder 1995; Raymont 2014). Vertical migrations were even found within the sand layer of the intertidal zone where *Euglena deses* shows diurnal patterns staining the top sand layer green when the cells move to the surface (Taylor 1967).

A direct proof that *Euglena* and *Paramecium* have the capacity to use gravity for orientation was obtained in a space experiment (Häder et al. 1990; Hemmersbach-Krause et al. 1993). Cell cultures were subjected to a parabolic flight on a TEXUS rocket (technical experiments under microgravity) which provides microgravity for about 6 min (Hemmersbach-Krause et al. 1993; Häder et al. 2010). Video downlink allowed online observation of the swimming cells and they were found to move in random directions, while the 1-g ground controls and post-flight analysis of flight samples showed a precise negative gravitaxis.

It is not amazing that all investigated motile flagellate and ciliate species have been found to respond to the gravitational field of the Earth since this force is invariable and has affected the evolution of the biota from the very beginning. However, the identification of the underlying mechanisms gave rise to many discussions and hypotheses. Today, by studies of gravitactic organisms on ground and in space, we obtained new insights into the cellular physiology. Especially asking the question, on which level of organization gravity impacts biological systems, motile unicellular organisms allowed exciting insights into gravity-related signaling pathways.

3.2   Gravitaxis and its Ecological Advantages

Many motile microorganisms demonstrate swimming parallel to the gravity vector of the Earth either upwards (negative gravitaxis) or downwards (positive gravitaxis) in order to optimize their position in the water column (Richter et al. 2007). Negative gravitaxis brings them close to the water surface which is of an advantage for photosynthetic organisms. Swimming downwards, e.g. into the sediment fits with their metabolic requirements: e.g., a direct correlation between the abundance of oxygen and gravitactic orientation has been demonstrated for the ciliate *Loxodes* and the flagellate *Euglena*. The microaerophilic *Loxodes* can be found predominantly in the sediment of lakes which it reaches by positive gravitaxis (Fenchel and Finlay 1984, 1990; Hemmersbach and Häder 1999) while the photosynthetic flagellate *Euglena gracilis* swims preferentially upwards (Lebert and Häder 1996; Fig. 3.1), thereby reaching its ecological niche.

An additional advantage of oriented movements with respect to the gravity vector is protection from detrimental stress factors such as solar UV-B (280–315 nm) radiation (Häder et al. 1999); when exposed to excessive visible or UV radiation some flagellates swim actively downward (Ntefidou et al. 2002; Richter et al. 2002b) or stop swimming and sediment passively out of the danger zone (Häder and Liu 1990; Sebastian et al. 1994). In *Euglena* the reversal of movement direction has

Fig. 3.1 Circular
histograms of positive
gravitactic orientation in the
ciliate *Loxodes striatus* (**a**)
and negative gravitactic
orientation in the flagellate
Euglena gracilis (**b**). The
lengths of each sector
indicate the relative number
of cells swimming in the
corresponding direction.
Redrawn after (Lebert and
Häder 1996; Hemmersbach
and Häder 1999)

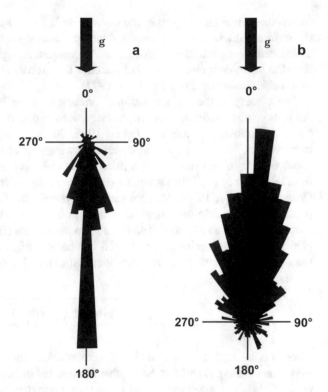

been attributed to the production of reactive oxygen species (ROS) as shown by
the fluorescent probe 2′,7′-dichlorodihydrofluorescein diacetate (DCFA-DAL) and
the application of ROS scavengers (Richter et al. 2003b, c). Also the application of
15 g/L NaCl reversed the direction of negative gravitaxis into a positive one (Richter
et al. 2003a).

The precision of gravitactic orientation can be modulated by other environmental
factors, such as oxygen concentration: *Loxodes* shows a precise positive gravitaxis at
high O_2 concentrations and is less oriented at low concentrations (Finlay et al. 1993).
In contrast, *Paramecium* shows negative gravitaxis (Hemmersbach-Krause and
Häder 1990; Hemmersbach and Donath 1995) which is more pronounced at a low
oxygen tension than at a higher one (Hemmersbach-Krause et al. 1991). Also the
feeding status and temperature modify the direction and precision of gravitaxis in
Paramecium (Moore 1903). As some microorganisms, such as unicellular algae
(flagellates) orient with respect to light as well as to gravity, these two kinds of
stimuli might be competing environmental clues since gravitaxis was weak in strong
light while in low light the gravity signal outcompetes light (Wager 1911). When
phototactic organisms are exposed to light under microgravity conditions their
phototaxis is more precise than on Earth, clearly showing that these two stimuli
operate synergistically under normal gravity conditions (Häder 1997).

Another advantage of gravitactic orientation is found in gametes which move to the surface and have a higher chance of meeting their sexual partner than when being distributed throughout the water column. In contrast, zoospores of many algae were found to swim downward in order to anchor themselves at the bottom and to grow into multicellular thalli (Callow et al. 1997).

While early studies of gravitational orientation were based on microscopic or population observations, modern investigations are carried out using computer-based cell tracking (Häder and Lebert 1985, 2000; Häder and Vogel 1991). The movements of orienting cells in a vertical swimming chamber are recorded by a video camera in real time using a microscope objective in an infrared monitoring light in order not to disturb gravitactic orientation by an interference of phototaxis (Vogel and Häder 1990). Modern tracking systems can follow numerous cells in parallel and calculate movement direction, swimming velocity and percentage of motile organisms as well as cell size and form parameters (Häder and Erzinger 2015; Häder 2017). The precision of gravitaxis can be quantified by a statistical approach: the r-value runs between 0 (random orientation) and 1 (all organisms move in the same direction).

$$r = \frac{\sqrt{\left(\sum \sin \alpha\right)^2 + \left(\sum \cos \alpha\right)^2}}{n} \tag{3.1}$$

where α is the angular deviation from the stimulus direction (in gravitaxis away from the center of gravity) and n is the number of recorded tracks (Batschelet 1981; Häder et al. 2005a). The mean movement angle θ of a population can be determined by

$$\theta = \arctan\left(\frac{\sum \sin \alpha}{\sum \cos \alpha}\right) \tag{3.2}$$

3.3 Gravitaxis—The Underlying Mechanism

Gravitaxis of motile microorganisms has been studied for a long time starting with experiments where swimming cells were observed in vertical glass tubes (Platt 1899; Köhler 1921). Earlier studies were descriptive but initiated a long-lasting and some-times emotional debate on the underlying mechanism. Two groups were formed, either postulating a pure physical mechanism or a physiological one reviewed e.g. by (Haupt 1962; Machemer and Bräucker 1992; Hemmersbach et al. 1999).

During the studies of gravitactic orientation a few model organisms have crys-tallized including the flagellates *Euglena, Chlamydomonas* and the ciliates *Bursaria, Paramecium, Stylonychia* and *Loxodes* because they allow studying gravity-signal transduction chains in unicellular organisms gathering several aspects, which will be exemplarily presented in the subsequent paragraphs.

3.4   Mechanisms of Gravity Perception Resulting in Gravitaxis

In Chap. 1 of this volume two principles for graviperception in single cells are being discussed, based on sedimenting masses. One is based on intracellular statoliths and the other on the force exerted by the whole mass of the cell itself. Both mechanisms require the existence of a sensitive structure that transforms the signal and transfers it to a motile apparatus and thus elicits a response. In gravitactically moving organisms, we can distinguish between the two postulated mechanisms of graviperception by a simple experimental approach, using media of different densities (Fig. 3.2). If a cell detects the gravivector by a heavy statolith which presses onto a sensor underneath this will work independently of the density of the medium, surrounding the cell. In contrast, if the gravireceptor detects the force exerted by sedimentation of the

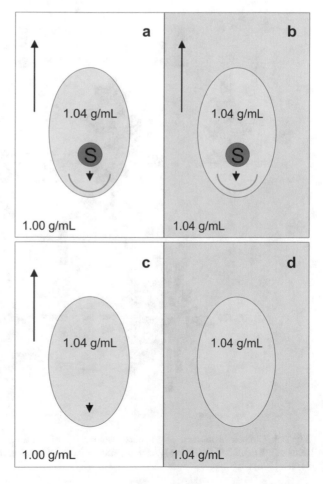

Fig. 3.2 Models for gravisensing. If gravitaxis is mediated by a statolith the cell would be capable of orienting in a medium of a density of 1.00 g/mL (**a**). It would also orient in a medium with increased density 1.04 g/mL (**b**) since the statolith still presses on a receptive structure. If in contrast the cell content with an assumed specific density of 1.04 g/mL presses onto a receptor, the cell would be able to orient in an outer medium of 1.00 g/mL (**c**), but not in a medium with a specific density of 1.04 g/mL (**d**), since the inner and outer medium are in equilibrium (Hemmersbach and Häder 1999)

whole cytoplasmic contents onto the cell membrane, this will only function when the outer medium has a specific density which is smaller than that of the cytoplasm. This principle cannot work, if the density of the outer medium and the one of the cytoplasm are equal, resulting in a loss of graviorientation. If the density of the outer medium exceeds that of the cytoplasm the cells will orient themselves in the opposite direction. The latter behavior was found in *Euglena* suspended in Ficoll solutions of increasing densities (Fig. 3.3; Lebert and Häder 1996). At low Ficoll concentrations the cells showed negative gravitaxis, at a concentration where the specific density equaled that of the cytoplasm the cells swam randomly and at higher concentrations they showed positive gravitaxis. In contrast, gravitaxis of *Loxodes,* which has intracellular statocyst-like organelles (heavy bodies of barium sulfate, see below), persisted even if the density difference between cell and medium was equal (Hemmersbach et al. 1998; Neugebauer et al. 1998).

The specific density of cells can be measured by isopycnic centrifugation using layers with increasing Ficoll concentrations (and thus increasing densities). For

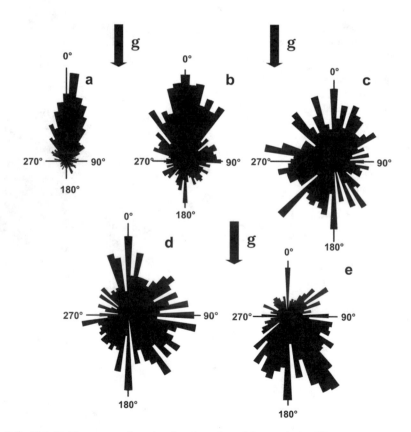

Fig. 3.3 Circular histograms of gravitactic orientation of *E. gracilis* in different Ficoll concentrations of 0% (**a**), 2.5% (**b**), 5% (**c**), 7.5% (**d**) and 10% (**e**). 0° indicates upwards direction (Lebert and Häder 1996)

Euglena two bands were found corresponding to specific densities of 1.046 and 1.054 g/mL; the lower value refers to young cells after cell division and the higher to older cells (Lebert et al. 1999b). After keeping a *Euglena* culture for more than 600 days completely enclosed the cell density had dropped to 1.011 g/mL and these cells had lost their ability for gravitactic orientation (Häder et al. 2005a). For comparison the ciliates *Bursaria truncatella* and *Paramecium caudatum* were found to have a specific density of 1.04 g/mL (Krause 1999), *Loxodes striatus* of 1.03 g/mL, while its Müller bodies consisting of barium sulfate and acting as statoliths have a specific density of 4.4 g/mL (Hemmersbach et al. 1998).

Several hypotheses have been formulated to explain the mechanism of gravitaxis (Machemer and Bräucker 1992; Barlow 1995; Hemmersbach et al. 1999). One interpretation posits a purely passive phenomenon based on buoyancy. In a tail-heavy cell the front end will point upward; in *Euglena* the apical trailing flagellum would propel the cell away from the center of gravity (Fukui and Asai 1985). However, microscopy does not reveal any heavy particles in the rear part of the cell. In order to test the hypothesis, we immobilized cells by injecting them into liquid nitrogen. These dead cells showed random orientation in the medium while the living controls were oriented with their apical ends upwards (Häder et al. 2005c). Roberts (1970) proposed an alternative model for graviperception in cells which lack sedimenting statoliths, dubbed drag-gravity model. While in the gravity-buoyancy model one Reynolds number describes sedimentation, in the drag-gravity model separate Reynolds numbers are used for the front and rear ends. In this explanation an elongated cell is envisioned as two coupled spheres. Stokes's law tells us that the larger rear end sediments faster than the smaller front end even though the two parts have the same specific density.

$$v = \frac{2(\rho_b - \rho_m)gr^2}{9\eta} \tag{3.3}$$

where v = velocity, ρ_b = specific density of the body, ρ_m = specific density of the medium, g = acceleration (9.81 m s^{-2}), r = radius and η = viscosity of the medium.

The propulsion-gravity model is based on a helical path during forward swimming which is found in many unicellular organisms where the long cell axis describes a cone and the front end rotates on a larger radius than the rear one (Winet and Jahn 1974). This is the result of the distance between the center of effort exerted by the flagella or cilia and the geometric center of the cell. This generates a torque in horizontally swimming cells, since the center of effort is closer to the front end than the geometric center. Viscosity of the medium counters sedimentation at low Reynolds numbers and turns the front end upwards. This model should apply better to *Euglena* with a single apical flagellum than to *Paramecium* where the cilia cover the whole cell body. Another hypothesis explains gravitaxis by a torque produced by the swimming cells (Kessler and Hill 1997). In contrast to the hypotheses based on passive orientation of the cell, it has been proposed that an active gravireceptor aligns the cell parallel to the gravity vector which results in a controlled steering movement (Dennison and Shropshire 1984).

Several environmental factors influence the precision of graviorientation mechanism. The precision of gravitaxis follows a circadian rhythm entrained by a light/dark rhythm (Lebert et al. 1999a; Nasir et al. 2014), which cannot be explained by a pure passive mechanism. In the dark phase the cells show a minimum in the precision of gravitactic orientation, an increase before the beginning of the light phase and a peak in the early afternoon (Häder and Lebert 2001). The rhythm persists even when the culture is transferred to constant light conditions (Lebert et al. 1999a). In parallel, the cells are more elongated during daytime while they are more rounded at night. An interesting phenomenon is that young *Euglena* after cell division show positive gravitaxis while older cells in the stationary phase display a pronounced negative gravitaxis (Stallwitz and Häder 1994). One explanation could be that the paramylon concentration in young cells is smaller than in older ones and thus the cells have a lower specific density. However, it was surprising when we found that the precise downward swimming of young cells could be reversed into an upward swimming upon application of heavy metal ions such as cadmium, copper, mercury or lead.

Application of a red background light strongly increases the precision of gravitaxis in the flagellate *Chlamydomonas* (Sineshchekov et al. 2000). The action spectrum of this response indicates the involvement of chlamy-rhodopsin which is the presumed receptor for phototaxis in this organism (Kianianmomeni and Hallmann 2014). Reducing the concentration of calcium ions in the medium even further enhances gravitaxis.

Ciliates have been called "swimming nerve cells" due to the electromotoric coupling of the membrane potential to ciliary beating. Like *Euglena, Paramecium* uses it's whole body mass for graviperception which sediments onto the lower cell membrane and activates mechanosensitive ion channels. Electrophysiological studies have revealed a polar distribution of mechanosensitive calcium and potassium ion channels, which upon stimulation either de- or hyperpolarize the cell membrane and in turn determine the ciliary activity and thus orientation and swimming speed of the cell. This polar distribution is a prerequisite for another gravity-related response of ciliates. Besides for orientation, gravity can also be used to modify the swimming speed. Speeding up during upward swimming (activation of potassium channels resulting in hyperpolarization) and decreasing during downward swimming (activation of calcium channels resulting in depolarization) enables the cells to compensate sedimentation (Machemer 1994; Gebauer et al. 1999).

A true gravireceptor potential was found in the ciliate *Stylonychia mytilus* proving that the above hypothesis is valid for the gravitactic transduction chain. Membrane potential changes of 4 mV after cell reorientation in the gravity field by turning the cell upside down identified the existence of gravireceptor potentials and thus support the hypothesis that the cytoplasm can exert the pressure onto a gravireceptor (Krause et al. 2010). The polar distribution of mechanosensitive ion channels and the signaling cascade between their stimulation and ciliary beating are prerequisites for using gravity for active speed control.

Representatives of the family Loxodidae kindled the interest in the search of gravisensing mechanisms in single cells. So-called Müller organelles, of which

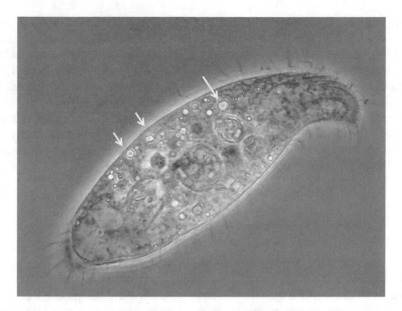

Fig. 3.4 Light microscopic image of *Loxodes striatus* showing the position of the Müller organelles, acting as statocyst-like organelle (courtesy K. Hausmann)

3–4 per cell can be found in *Loxodes striatus* (Fenchel and Finlay 1986) bear all characteristics of true gravisensors (Fig. 3.4). A heavy mass of barium sulfate within a vacuole is fixed to a ciliary stick. This configuration links spatial orientation of the cell and thus the statolith movements to changes in the membrane potential, which determines ciliary activity and thus swimming direction. So far, *Loxodes* represents the only animal cell type in which the existence and function of a true cellular gravireceptor could be demonstrated. Destruction of the connectivity between the Müller body and the ciliary stick by means of a laser beam resulted in a loss of gravitactic orientation, while control shots in the nearby cytoplasm had no effect.

3.5   Thresholds Characterize Gravireceptors

The residual acceleration during a TEXUS rocket flight is on the order of 10^{-3} to 10^{-4} g. Since the cells do not respond to this force the threshold of *Euglena, Loxodes* and *Paramecium* must be higher. In order to measure their threshold for gravitaxis, experiments were carried out on board the Space Shuttle Columbia during the second International Microgravity mission (IML-2). The cell suspensions were transferred onto a slow rotating centrifuge microscope (NiZeMi; Friedrich et al. 1996) and subjected to increasing accelerations between 0 and 1.5 g. The analysis of the data showed that the threshold for gravitaxis in *Euglena* is ≤ 0.16 g and saturation was found at 0.64 g (Häder et al. 1996). *Paramecium* showed a similar threshold for graviperception between 0.16 and 0.3 g (Hemmersbach et al. 1996). Using a centrifuge with an attached microscope on TEXUS and MAXUS sounding rockets, providing accelerations during 6 min and 13 min of microgravity the threshold could be measured even more precisely to be ≤0.12 g for *Euglena* (Häder et al. 1997). The direct comparison between *Paramecium* and *Loxodes* confirmed the threshold for *Paramecium* in the range of 0.16 and 0.3 g and revealed higher sensitivity of ≤0.15 g for *Loxodes*. The swimming velocity of *Euglena* followed precisely Stokes' law for sedimentation (Häder et al. 1996) in contrast to the ciliates *Paramecium* and *Loxodes* which show gravikinesis (Machemer-Röhnisch et al. 1998). Also no adaptation to the microgravity conditions was found.

3.6   Sensory Transduction Chain for Gravitaxis

If it is correct that the heavier cellular cytoplasm presses onto the lower membrane, the next question is what is the nature of the gravireceptor which perceives this force (Häder et al. 2005b). One option could be mechanosensitive ion channels which can be found in many organisms from bacteria to vertebrates. These membrane-bound proteins are operated by small mechanical forces and gate the influx of cations (Jarman and Groves 2013). In order to be effective first an ion gradient has to be established across the outer membrane. In fact, in *Euglena* an ATPase was found which pumps Ca^{2+} into the outer medium and generates a gradient of six orders of magnitude. This pump can be blocked by vanadate and as a result gravitactic orientation is impaired (Lebert et al. 1997). Likewise, the synthetic ionophore calcimycin (A23187) can be integrated into the cell membrane, which breaks down the Ca^{2+} gradient and reduces gravitactic orientation (Lebert et al. 1996). The presumed mechanosensitive ion channels can be blocked by gadolinium, which hinders the Ca^{2+} influx into the cell during gravitactic stimulation (Häder and Hemmersbach 1997). Likewise, application of potassium or cadmium strongly reduce the precision of gravitaxis in *Euglena* (Lebert et al. 1996). Also in *Chlamydomonas* mechanosensitive ion channels are thought to be involved in graviperception (Yoshimura 2011) since mutants which lack these proteins did not show gravitaxis (Häder et al. 2006).

After stimulation the Ca^{2+} gated by the channel protein can be visualized using a fluorescent chromophore such as Calcium Crimson previously loaded into the cells by electroporation; a strongly fluorescent signal could be detected near the front end of the cell. This was confirmed in the colorless *E. longa* and the likewise colorless, but gravitactic competent mutants 1F, 9F and FB (Häder and Lebert 2001). The fluorescent signal could also be detected during hypergravity phases of parabolic airplane maneuvers (Richter et al. 2002c) as well as on a centrifuge during a parabolic rocket flight when the threshold was exceeded (Häder and Lebert 1998). Gadolinium, which blocks stretch-sensitive ion channels (Sachs and Morris 1998) impaired the signal. The mechanosensitive ion channels are supposed to be located at the front end of the cell adjacent to the trailing flagellum. In that position they are activated when the flagellum is pointing downwards during the helical rotation of the cell (Häder et al. 2005a). Repetitive activation of the channels results in a bending of the flagellum mediating upward steering. The influx of Ca^{2+} ions during gravitactic orientation results in a change in the membrane potential which can be demonstrated using the electrochromic absorbance band shift of Oxonol VI incorporated into the membrane (Richter et al. 2001) showing that during reorientation of the cells the membrane potential changed accordingly (Richter et al. 2006). Disturbing the membrane potential by application of the lipophilic cation $TPMP^+$ (triphenylmethyl phosphonium) reduced the precision of gravitactic orientation in *Euglena* (Lebert et al. 1996).

The identification of the mechanosensitive ion channel responsible as gravi-receptor in *Euglena* was revealed by molecular biology. Using several primers against the conserved regions of the corresponding gene in *Saccharomyces* (Häder et al. 2003) yielded more than 1500 PCR products which were cloned in plasmids and sequenced (Häder et al. 2009). Most of these genes coded for proteins with different functions, but some resembled genes for a TRP (transient receptor potential-like) channel. The large TRP protein family is involved in photoperception, nociception, thermosensation, mechanosensing, taste, osmolar sensation, and fluid flow detection (Nilius et al. 2012). The method of RNA interference (RNAi) was used to identify the specific TRP involved in graviperception. For this purpose, double-stranded RNA fragments (19–23 nucleotides) are introduced into the cell which causes a sequence-specific post-transcriptional gene silencing. The RNA fragment attaches itself to the mRNA coding for the protein in question and thus blocks its translation and the synthesis of the corresponding protein. Four PCR transcripts were found in *Euglena* using degenerated primers reflecting the pore region of the mechanosensitive channel. Using RNAi showed that three of these products were not involved in gravitaxis, but inhibition of the fourth (TRPC7) blocked graviorientation for up to 30 days (Fig. 3.5).

The Ca^{2+} ions gated into the cytoplasm during gravistimulation could bind to calmodulins, a group of proteins with usually four binding sites for Ca^{2+}, which function as universal messenger in many organisms from bacteria to vertebrates (Adler 2013). Calmodulin can be inhibited by trifluoperazine, fluphenazine or W7 (Naccache 1985; Russo et al. 2014; Son et al. 2014). Application of these drugs to *Euglena* strongly impaired gravitaxis (Häder et al. 2006). In order to test the hypothesis that a calmodulin is involved in the gravitaxis sensory transduction chain we searched for corresponding genes in *Euglena*. Surprisingly we found and sequenced five genes (CaM.1–CaM.5; Daiker et al. 2010). One of these (CaM.1) was already known in *E. gracilis* and shown to be located under the pellicula (Toda et al. 1992). Applying again RNAi we blocked the protein synthesis of each calmodulin in separate populations. Using RT-PCR confirmed that blocking the individual mRNA suppressed the synthesis of the corresponding calmodulin. Inhibiting CaM.1 caused strong abnormalities of the cell form and inhibited swimming motility even though the flagellum was visible. RNAi against CaM.3–CaM.5 did not affect gravi-orientation but inhibition of CaM.2 effectively impaired gravitaxis.

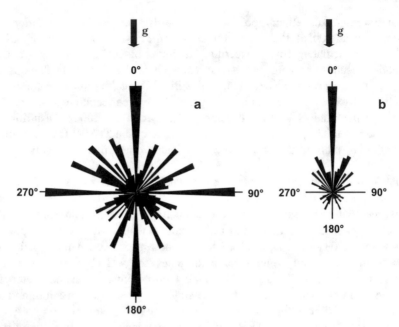

Fig. 3.5 (a) Circular histograms of gravitactic orientation in *E. gracilis* 15 days after RNAi against TRPC7, (b) control. Redrawn after (Häder et al. 2009)

Previously it had been shown that a light-activated adenylyl cyclase (PAC) is involved in the photoperception and phototaxis in *Euglena* which produces cyclic adenosine monophosphate (cAMP) from ATP when excited by light (Iseki et al. 2002; Häder and Iseki 2017). Since earlier experiments had shown that the intracellular cAMP concentration increases in parallel with the precision of gravitaxis during the circadian rhythm (Lebert et al. 1999a) we designed a very complex space experiment on TEXUS 36 (Tahedl et al. 1998). 2-mL syringes filled with 1 mL *Euglena* suspension each were connected by tubes with corresponding syringes containing 1 mL ethanol but separated by rubber balls. 112 syringe assemblies were mounted on a centrifuge at different radii to produce different accelerations. At predetermined times during the flight in separate groups of 8–12 syringes each the ethanol was injected into the cell populations by hydraulic pistons to chemically fix cells at different times and accelerations. After retrieval of the payload the cAMP of the cells was quantified with a radioimmunoassay in a scintillation counter. At microgravity and at 0.08 g (which is below the gravitactic threshold) very low cAMP concentrations were found, but at accelerations of 0.12 and 0.16 the concentration increased significantly (Tahedl et al. 1998). In a parallel experiment gadolinium-treated (1 mM) cells showed a much lower increase in cAMP following gravistimulation. Caffeine-treated (10 mM) cells had no elevated cAMP concentrations during gravistimulation but had about three times higher cAMP concentrations in microgravity (Tahedl et al. 1998). The application of indomethacin, a known inhibitor of the adenylyl cyclase (Wang et al. 2012), impairs gravitaxis (Richter et al.

2002a; Häder et al. 2006) while forskolin activates the adenylyl cyclase (Schwer et al. 2013) and increases the precision of gravitaxis in *Euglena* (Häder et al. 2006). Phosphodiesterase quenches the cAMP after stimulation (Cai et al. 2015). Caffeine, theophylline and IBMX are known inhibitors of this enzyme (Cameron and Baillie 2012; Steck et al. 2014), so that the cAMP signal stays high after application and the gravitactic activity increases (Lebert et al. 1997; Häder and Lebert 2002; Streb et al. 2002). Application of 8-bromo-cAMP, incorporated into the cells, functions as a cAMP analog, but is not quenched by the phosphodiesterase. As a result the precision of gravitaxis increased (Lebert et al. 1997). cAMP seems to be a universal component of sensory transduction chains since its involvement has been found in the control of movement and development in the slime mold *Dictyostelium* (Renart et al. 1981; Sultana et al. 2012) and in the gravitactic sensory transduction chain of *Paramecium* (Bräucker et al. 2001; Hemmersbach and Braun 2006).

The next question is how the cAMP signal controls the flagellar activity which is instrumental in the steering response of the cell. Favaro et al. (2012) had shown that cAMP can activate a protein kinase A. Staurosporine, which is known to be an inhibitor of protein kinases (Chang and Kaufman 2000) impairs gravitaxis in *Euglena* (Häder et al. 2010). But it is interesting that prolonged exposure (225 min) to the drug reversed the negative gravitaxis into a positive one as did strong visible and UV radiation or salt stress (see above). This drug also impairs negative phototaxis which could indicate that both sensory transduction chains share the same final step being the activation of a phosphokinase A (Häder and Iseki 2017).

Using the same molecular biological tools involving degenerate primers as described above showed that *Euglena* possesses at least five isoforms of protein kinase A (PK.1–PK.5). The full range of the RNA sequence was revealed by RACE-PCR. RNAi against the different isoforms showed that PK.1–PK.3 and PK.5 are not involved in the gravitactic sensory transduction chain (Daiker et al. 2011). In contrast, inhibition of PK.4 effectively blocked gravitaxis for several weeks. Three weeks after the RNAi treatment a positive gravitaxis was observed. This finding parallels the result by staurosporine inhibition. In addition, RNAi of PK.4 blocked phototaxis which further confirms that gravitaxis and phototaxis share the same final step, the activation of a protein kinase A by cAMP.

3.7   Conclusions and Open Questions

Gravitactic unicellular organisms are specialized for gravity sensing. They use this information for their spatial orientation (gravitaxis). Summarizing the findings allows us to construct a complete gravitaxis signal transduction chain, such as in *Euglena* (Fig. 3.6). During forward locomotion the cell rotates around its horizontal axis at about 1 Hz. When the flagellum points downwards the content of the cell presses onto the lower membrane activating the TRP channels, thought to be located at the front end adjacent to the flagellum, which allows Ca^{2+} to enter the cell along a previously established gradient by a Ca-ATPase. In fact, several rotations are

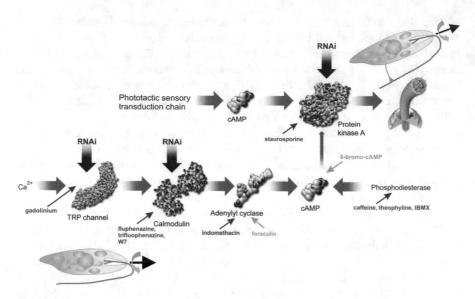

Fig. 3.6 Model of the gravitaxis sensory transduction chain in *Euglena*. Elements proven by RNAi are indicated. Inhibitors are shown in red and activators in blue. The phototaxis sensory transduction chain shares the same final element, the protein kinase A

necessary to increase the Ca^{2+} concentration above a threshold necessary for the transduction chain to operate and to swing out the flagellum. The Ca^{2+} entering the cell binds to a specific calmodulin which in turn activates an adenylyl cyclase. This produced cAMP in turn activates a protein kinase A before it is quenched by a phosphodiesterase. The protein kinase A is probably located inside the flagellum where it instrumentalizes the bending of the axoneme by protein phosphorylation resulting in a course correction of the cell path.

These results show that *Euglena* uses an active gravireceptor and a complex transduction chain and is not passively aligned in the water column. Open questions include how the cell switches from positive to negative gravitaxis. ROS seem to be involved in this control (Richter et al. 2003c). Key elements in the gravisensory transduction chain such as mechanosensitive ion channels, cAMP and reorientation of cilia have been found also in ciliates. Thus, it is intriguing to posit that similar molecular mechanisms are involved in their graviorientation. To identify whether these mechanisms and capacities are used by migrating cells—also in the human body—will be a challenge of the future.

References

Adler EM (2013) Bacteria under pressure, calcium channel internalization, and why cockroaches avoid glucose-baited traps. J Gen Physiol 142:1–2

Barlow PW (1995) Gravity perception in plants: a multiplicity of systems derived by evolution? Plant Cell Environ 18:951–962

Batschelet E (1981) Circular statistics in biology. Academic Press, London

Bräucker R, Cogoli A, Hemmersbach R (2001) Graviperception and graviresponse at the cellular level. In: Baumstark-Khan C, Horneck G (eds) Astrobiology: the quest for the conditions of life. Springer Verlag, Berlin, pp 284–297

Cai Y, Nagel DJ, Zhou Q, Cygnar KD, Zhao H, Li F, Pi X, Knight PA, Yan C (2015) Role of cAMP-phosphodiesterase 1C signaling in regulating growth factor receptor stability, vascular smooth muscle cell growth, migration, and neointimal hyperplasia. Circ Res 116:1120–1132

Callow ME, Callow JA, Pickett-Heaps JD, Wetherbee R (1997) Primary adhesion of Enteromorpha (Chlorophyta, Ulvales) propagules: quantitative settlement studies and video microscopy. J Phycol 33:938–947

Cameron R, Baillie GS (2012) cAMP-specific phosphodiesterases: modulation, inhibition, and activation. In: Botana LM, Loza M (eds) Therapeutic targets: modulation, inhibition, and activation. Wiley, Hoboken, NJ, p 1

Chang SC, Kaufman PB (2000) Effects of staurosporine, okadaic acid and sodium fluoride on protein phophorylation in graviresponding oat shoot pulvini. Plant Physiol Biochem 38:315–323

Clegg MR, Maberly SC, Jones RI (2003) Chemosensory behavioural response of freshwater phytoplanktonic flagellates. Plant Cell Environ 27:123–135

Daiker V, Häder D-P, Lebert M (2010) Molecular characterization of calmodulins involved in the signal transduction chain of gravitaxis in Euglena. Planta 231:1229–1236

Daiker V, Häder D-P, Richter RP, Lebert M (2011) The involvement of a protein kinase in phototaxis and gravitaxis of Euglena gracilis. Planta 233:1055–1062

Dennison DS, Shropshire W Jr (1984) The gravireceptor of Phycomyces. Its development following gravity exposure. J Gen Physiol 84:845–859

Eggersdorfer B, Häder D-P (1991) Phototaxis, gravitaxis and vertical migrations in the marine dinoflagellate Prorocentrum micans. FEMS Microbiol Ecol 85:319–326

Favaro E, Granata R, Miceli I, Baragli A, Settanni F, Perin PC, Ghigo E, Camussi G, Zanone M (2012) The ghrelin gene products and exendin-4 promote survival of human pancreatic islet endothelial cells in hyperglycaemic conditions, through phosphoinositide 3-kinase/Akt, extra-cellular signal-related kinase (ERK) 1/2 and cAMP/protein kinase a (PKA) signalling pathways. Diabetologia 55:1058–1070

Fenchel T, Finlay BJ (1984) Geotaxis in the ciliated protozoon Loxodes. J Exp Biol 110:17–33

Fenchel T, Finlay BJ (1986) The structure and function of Müller vesicles in loxodid ciliates. J Protozool 33:69–76

Fenchel T, Finlay BJ (1990) Oxygen toxicity, respiration and behavioural responses to oxygen in free-living anaerobic ciliates. J Gen Microbiol 136:1953–1959

Fiedler B, Börner T, Wilde A (2005) Phototaxis in the cyanobacterium Synechocystis sp. PCC 6803: role of different photoreceptors. Photochem Photobiol 81:1481–1488

Finlay BJ, Tellez C, Esteban G (1993) Diversity of free-living ciliates in the sandy sediment of a Spanish stream in winter. J Gen Microbiol 139:2855–2863

Fraenkel GS, Gunn DL (1961) The orientation of animals (Kineses, taxes and compass reactions). Dover Publication, New York

Friedrich ULD, Joop O, Pütz C, Willich G (1996) The slow rotating centrifuge microscope NIZEMI – a versatile instrument for terrestrial hypergravity and space microgravity research in biology and materials science. J Biotechnol 47:225–238

Fukui K, Asai H (1985) Negative geotactic behavior of Paramecium caudatum is completely described by the mechanism of buoyancy-oriented upward swimming. Biophys J 47:479–482

Gebauer M, Watzke D, Machemer H (1999) The gravikinetic response of Paramecium is based on orientation-dependent mechanotransduction. Naturwissenschaften 86:352–356

Häder D-P (1991) Strategy of orientation in flagellates. In: Riklis E (ed) Photobiology. Springer, Boston, MA, pp 497–510

Häder D-P (1997) Gravitaxis and phototaxis in the flagellate Euglena studied on TEXUS missions. In: Cogoli A, Friedrich U, Mesland D, Demets R (eds) Life science experiments performed on sounding rockets (1985–1994). ESA Publications Division, Noordwijk, pp 77–79

Häder D-P (2017) Image analysis for bioassays – the basics. In: Häder D-P, Erzinger GS (eds) Bioassays: advanced methods and applications. Elsevier, Atlanta, GA, pp 69–98

Häder D-P, Erzinger GS (2015) Advanced methods in image analysis as potent tools in online biomonitoring of water resources. Recent Pat Top Imaging 5:112–118

Häder D-P, Hemmersbach R (1997) Graviperception and graviorientation in flagellates. Planta 203:7–10

Häder D-P, Iseki M (2017) Photomovement in *Euglena*. In: Schwartzbach S, Shigeoka S (eds) *Euglena*: biochemistry, cell and molecular biology. Springer, Cham, pp 207–235

Häder D-P, Lebert M (1985) Real time computer-controlled tracking of motile microorganisms. Photochem Photobiol 42:509–514

Häder D-P, Lebert M (1998) Mechanism of gravitactic signal perception and signal transduction of *Euglena gracilis*. Micrograv News 11:14

Häder D-P, Lebert M (2000) Real-time tracking of microorganisms. In: Häder D-P (ed) Image analysis: methods and applications. CRC Press, Boca Raton, pp 393–422

Häder D-P, Lebert M (2001) Graviperception and gravitaxis in algae. Adv Space Res 27:861–870

Häder D-P, Lebert M (2002) Graviorientation in flagellates. Proceedings 2nd China-Germany workshop on microgravity sciences, National Microgravity Laboratory, Chinese Academy of Sciences, Dunhuang, Beijing, China, September 1–3, 2002

Häder D-P, Liu SM (1990) Effects of artificial and solar UV-B radiation on the gravitactic orientation of the dinoflagellate, *Peridinium gatunense*. FEMS Microbiol Ecol 73:331–338

Häder D-P, Vogel K (1991) Simultaneous tracking of flagellates in real time by image analysis. J Math Biol 30:63–72

Häder D-P, Vogel K, Schäfer J (1990) Responses of the photosynthetic flagellate, *Euglena gracilis*, to microgravity. Appl Micrograv Technol 3:110–116

Häder D-P, Rosum A, Schäfer J, Hemmersbach R (1996) Graviperception in the flagellate *Euglena gracilis* during a shuttle space flight. J Biotechnol 47:261–269

Häder D-P, Porst M, Tahedl H, Richter P, Lebert M (1997) Gravitactic orientation in the flagellate *Euglena gracilis*. Microgravity Sci Technol 10:53–57

Häder D-P, Lebert M, Richter P (1999) Gravitaxis and graviperception in flagellates and ciliates. Proceedings 14th ESA symposium on European rocket and balloon programmes and related research (ESA SP-437), Potsdam, Germany

Häder D-P, Lebert M, Richter P, Ntefidou M (2003) Gravitaxis and graviperception in flagellates. Adv Space Res 31:2181–2186

Häder D-P, Hemmersbach R, Lebert M (2005a) Gravity and the behavior of unicellular organisms. Cambridge University Press, Cambridge

Häder D-P, Richter P, Daiker V, Lebert M (2005b) Molecular transduction chain for graviorientation in flagellates. ELGRA News 24:74

Häder D-P, Richter P, Ntefidou M, Lebert M (2005c) Gravitational sensory transduction chain in flagellates. Adv Space Res 36:1182–1188

Häder D-P, Richter P, Lebert M (2006) Signal transduction in gravisensing of flagellates. Signal Transduct 6:422–431

Häder D-P, Richter P, Schuster M, Daiker V, Lebert M (2009) Molecular analysis of the graviperception signal transduction in the flagellate *Euglena gracilis*: involvement of a transient receptor potential-like channel and a calmodulin. Adv Space Res 43:1179–1184

Häder D-P, Faddoul J, Lebert M, Richter P, Schuster M, Richter R, Strauch SM, Daiker V, Sinha R, Sharma N (2010) Investigation of gravitaxis and phototaxis in *Euglena gracilis*. In: Sinha R, Sharma NK, Rai AK (eds) Advances in life sciences. IK International Publishing House, New Delhi, pp 117–131

Haupt W (1962) Geotaxis. In: Ruhland W (ed) Handbuch der Pflanzenphysiologie. Springer-Verlag, Berlin, pp 390–395

Hemmersbach R, Bräucker R (2002) Gravity-related behaviour in ciliates and flagellates. Adv Space Biol Med 8:59–75

Hemmersbach R, Braun M (2006) Gravity-sensing and gravity-related signaling pathways in unicellular model systems of protists and plants. Signal Transduct 6:432–442

Hemmersbach R, Donath R (1995) Gravitaxis of *Loxodes* and *Paramecium*. Eur J Protistol 31:433

Hemmersbach R, Häder D-P (1999) Graviresponses of certain ciliates and flagellates. FASEB J 13: S69–S75

Hemmersbach R, Voormanns R, Briegleb W, Rieder N, Häder D-P (1996) Influence of accelerations on the spatial orientation of *Loxodes* and *Paramecium*. J Biotechnol 47:271–278

Hemmersbach R, Voormanns R, Bromeis B, Schmidt N, Rabien H, Ivanova K (1998) Comparative studies of the graviresponses of *Paramecium* and *Loxodes*. Adv Space Res 21:1285–1289

Hemmersbach R, Volkmann D, Häder D-P (1999) Graviorientation in protists and plants. J Plant Physiol 154:1–15

Hemmersbach-Krause R, Häder D-P (1990) Negative gravitaxis (geotaxis) of *Paramecium* – demonstrated by image analysis. Appl Micrograv Technol 4:221–223

Hemmersbach-Krause R, Briegleb W, Häder D-P (1991) Dependence of gravitaxis in *Paramecium* on oxygen. Eur J Protistol 27:278–282

Hemmersbach-Krause R, Briegleb W, Häder D-P, Vogel K, Grothe D, Meyer I (1993) Orientation of *Paramecium* under the conditions of weightlessness. J Eukaryot Microbiol 40:439–446

Iseki M, Matsunaga S, Murakami A, Ohno K, Shiga K, Yoshida C, Sugai M, Takahashi T, Hori T, Watanabe M (2002) A blue-light-activated adenylyl cyclase mediates photoavoidance in *Euglena gracilis*. Nature 415:1047–1051

Jarman AP, Groves AK (2013) The role of atonal transcription factors in the development of mechanosensitive cells. Semin Cell Dev Biol 24:438–447

Kessler JO, Hill NA (1997) Complementarity of physics, biology and geometry in the dynamics of swimming micro-organisms. In: Physics of biological systems. Springer, Berlin, pp 325–340

Kianianmomeni A, Hallmann A (2014) Algal photoreceptors: in vivo functions and potential applications. Planta 239:1–26

Köhler O (1921) Über die Geotaxis von *Paramecium*. Verhandlungen der Deutschen Zoologischen Gesellschaft 26:69–71

Krause M (1999) Elektrophysiologie, Mechanosensitivität und Schwerkraftbeantwortung von *Bursaria truncatella* Diploma thesis, Fakultät für Biologie der Ruhr-Universität Bochum

Krause M, Bräucker R, Hemmersbach R (2010) Gravikinesis in *Stylonychia mytilus* is based on membrane potential changes. J Exp Biol 213:161–171

Lebert M, Häder D-P (1996) How *Euglena* tells up from down. Nature 379:590

Lebert M, Richter P, Porst M, Häder D-P (1996) Mechanism of gravitaxis in the flagellate *Euglena gracilis*. Proceedings of the 12th C.E.B.A.S.workshops. Annual issue 1996, Ruhr-University of Bochum, Bochum, Germany

Lebert M, Richter P, Häder D-P (1997) Signal perception and transduction of gravitaxis in the flagellate *Euglena gracilis*. J Plant Physiol 150:685–690

Lebert M, Porst M, Häder D-P (1999a) Circadian rhythm of gravitaxis in *Euglena gracilis*. J Plant Physiol 155:344–349

Lebert M, Porst M, Richter P, Häder D-P (1999b) Physical characterization of gravitaxis in *Euglena gracilis*. J Plant Physiol 155:338–343

Machemer H (1994) Gravity-dependent modulation of swimming rate in ciliates. Acta Protozool 33:53–57

Machemer H, Bräucker R (1992) Gravireception and graviresponses in ciliates. Acta Protozool 31:185–214

Machemer-Röhnisch S, Bräucker R, Machemer H (1998) Graviresponses of gliding and swimming *Loxodes* using step transition to weightlessness. J Eukaryot Microbiol 45:411–418

Maree AFM, Panfilov AV, Hogeweg P (1999) Migration and thermotaxis of *Dictyostelium discoideum* slugs, a model study. J Theor Biol 199:297–309

Moore A (1903) Some facts concerning geotropic gatherings of paramecia. Am J Physiol 9:238–244

Naccache PH (1985) Neutrophil activation and calmodulin antagonists. In: Hidaka H, Hartshorne DJ (eds) Calmodulin antagonists and cellular physiology. Academic Press, Orlando, pp 149–160

Nasir A, Strauch S, Becker I, Sperling A, Schuster M, Richter P, Weißkopf M, Ntefidou M, Daiker V, An Y (2014) The influence of microgravity on *Euglena gracilis* as studied on Shenzhou 8. Plant Biol 16:113–119

Neugebauer DC, Machemer-Röhnisch S, Nagel U, Bräucker R, Machemer H (1998) Evidence of central and peripheral gravireception in the ciliate *Loxodes striatus*. J Comp Physiol A 183:303–311

Nilius B, Appendino G, Owsianik G (2012) The transient receptor potential channel TRPA1: from gene to pathophysiology. Pflugers Arch 464:425–458

Ntefidou M, Richter P, Streb C, Lebert M, Häder D-P (2002) High light exposure leads to a sign change in gravitaxis of the flagellate *Euglena gracilis*. Life in space for life on earth. 8th European symposium on life sciences research in space. 23rd annual international gravitational physiology meeting, Karolinska Institutet, Stockholm, Sweden, ESA SP-501

Piazena H, Häder D-P (1995) Vertical distribution of phytoplankton in coastal waters and its detection by backscattering measurements. Photochem Photobiol 62:1027–1034

Platt JB (1899) On the specific gravity of *Spirostomum, Paramecium* and the tadpole in relation to the problem of geotaxis. Am Nat 33:31

Raymont JE (2014) Plankton & Productivity in the oceans: volume 1: phytoplankton. Pergamon Press, Oxford

Renart MF, Sebastian J, Mato JM (1981) Adenylate cyclase activity in permeabilized cells from *Dictyostelium discoideum*. Cell Biol Int Rep 5:1045–1054

Rhiel E, Häder D-P, Wehrmeyer W (1988) Diaphototaxis and gravitaxis in a freshwater *Cryptomonas*. Plant Cell Physiol 29:755–760

Richter P, Lebert M, Korn R, Häder D-P (2001) Possible involvement of the membrane potential in the gravitactic orientation of *Euglena gracilis*. J Plant Physiol 158:35–39

Richter P, Ntefidou M, Streb C, Lebert M, Häder D-P (2002a) Physiological characterization of gravitaxis in *Euglena gracilis*. J Gravit Physiol 9:279–280

Richter PR, Ntefidou M, Streb C, Faddoul J, Lebert M, Häder D-P (2002b) High light exposure leads to a sign change of gravitaxis in the flagellate *Euglena gracilis*. Acta Protozool 41:343–351

Richter PR, Schuster M, Wagner H, Lebert M, Häder D-P (2002c) Physiological parameters of gravitaxis in the flagellate *Euglena gracilis* obtained during a parabolic flight campaign. J Plant Physiol 159:181–190

Richter P, Börnig A, Streb C, Ntefidou M, Lebert M, Häder D-P (2003a) Effects of increased salinity on gravitaxis in *Euglena gracilis*. J Plant Physiol 160:651–656

Richter P, Ntefidou M, Streb C, Lebert M, Häder D-P (2003b) The role of reactive oxygen species (ROS) in signaling of light stress. Recent Res Dev Biochem 4:957–970

Richter PR, Streb C, Ntefidou M, Lebert M, Häder D-P (2003c) High light-induced sign change of gravitaxis in the flagellate *Euglena gracilis* is mediated by reactive oxygen species. Acta Protozool 42:197–204

Richter PR, Schuster M, Meyer I, Lebert M, Häder D-P (2006) Indications for acceleration-dependent changes of membrane potential in the flagellate *Euglena gracilis*. Protoplasma 229:101–108

Richter PR, Häder D-P, Gonçalves RJ, Marcoval MA, Villafañe VE, Helbling EW (2007) Vertical migration and motility responses in three marine phytoplankton species exposed to solar radiation. Photochem Photobiol 83:810–817

Roberts AM (1970) Geotaxis in motile micro-organisms. J Exp Biol 53:687–699

Russo E, Salzano M, De Falco V, Mian C, Barollo S, Secondo A, Bifulco M, Vitale M (2014) Calcium/calmodulin-dependent protein kinase II and its endogenous inhibitor α in medullary thyroid cancer. Clin Cancer Res 20:1513–1520

Sachs F, Morris CE (1998) Mechanosensitive ion channels in nonspecialized cells. In: Blaustein MP, Greger R, Grunicke H et al (eds) Reviews of physiology and biochemistry and pharmacology. Springer-Verlag, Berlin, pp 1–78

Schwer CI, Lehane C, Guelzow T, Zenker S, Strosing KM, Spassov S, Erxleben A, Heimrich B, Buerkle H, Humar M (2013) Thiopental inhibits global protein synthesis by repression of eukaryotic elongation factor 2 and protects from hypoxic neuronal cell death. PLoS One 8: e77258

Sebastian C, Scheuerlein R, Häder D-P (1994) Graviperception and motility of three *Prorocentrum* strains impaired by solar and artificial ultraviolet radiation. Mar Biol 120:1–7

Simmons SL, Sievert SM, Frankel RB, Bazylinski DA, Edwards KJ (2004) Spatiotemporal distribution of marine magnetotactic bacteria in a seasonally stratified coastal salt pond. Appl Environ Microbiol 70:6230–6239

Sineshchekov O, Lebert M, Häder D-P (2000) Effects of light on gravitaxis and velocity in *Chlamydomonas reinhardtii*. J Plant Physiol 157:247–254

Son YK, Li H, Jung ID, Park Y-M, Jung W-K, Kim HS, Choi I-W, Park WS (2014) The calmodulin inhibitor and antipsychotic drug trifluoperazine inhibits voltage-dependent K$^+$ channels in rabbit coronary arterial smooth muscle cells. Biochem Biophys Res Commun 443:321–325

Stallwitz E, Häder D-P (1994) Effects of heavy metals on motility and gravitactic orientation of the flagellate, *Euglena gracilis*. Eur J Protistol 30:18–24

Steck R, Hill S, Robison RA, O'Neill KL (2014) Pharmacological reversal of caffeine-mediated phagocytic suppression. Cancer Res 74:4861

Streb C, Richter P, Ntefidou M, Lebert M, Häder D-P (2002) Sensory transduction of gravitaxis in *Euglena gracilis*. J Plant Physiol 159:855–862

Sultana H, Neelakanta G, Rivero F, Blau-Wasser R, Schleicher M, Noegel AA (2012) Ectopic expression of cyclase associated protein CAP restores the streaming and aggregation defects of adenylyl cyclase a deficient *Dictyostelium discoideum* cells. BMC Dev Biol 12:3

Tahedl H, Richter P, Lebert M, Häder D-P (1998) cAMP is involved in gravitaxis signal transduction of *Euglena gracilis*. Microgravity Sci Technol 11:173–178

Taylor F (1967) The occurrence of *Euglena deses* on the sands of the Sierra Leone peninsula. J Ecol 55:345–359

Toda H, Yazawa M, Yagi K (1992) Amino acid sequence of calmodulin from *Euglena gracilis*. Eur J Biochem 205:653–660

Vogel K, Häder D-P (1990) Simultaneuos tracking of flagellates in real time by image analysis. Proceedings of the fourth European symposium on life science research in space (ESA SP-307)

Wadhams GH, Armitage JP (2004) Making sense of it all: bacterial chemotaxis. Nat Rev Mol Cell Biol 5:1024–1037

Wager H (1911) On the effect of gravity upon the movements and aggregation of *Euglena viridis*, Ehrb., and other micro-organisms. Philos Trans R Soc Lond B 201:333–390

Wang J, Sun Y, Tomura H, Okajima F (2012) Ovarian cancer G-protein-coupled receptor 1 induces the expression of the pain mediator prostaglandin E2 in response to an acidic extracellular environment in human osteoblast-like cells. Int J Biochem Cell Biol 44:1937–1941

Winet H, Jahn TL (1974) Geotaxis in protozoa: I. A propulsion-gravity model for *Tetrahymena* (Ciliata). J Theor Biol 46:449–465

Yoshimura K (2011) Stimulus perception and membrane excitation in unicellular alga *Chlamydomonas*. In: Coding and decoding of calcium signals in plants. Springer, Berlin, pp 79–91

Chapter 4
Gravitropism in Tip-Growing Rhizoids and Protonemata of Characean Algae

Markus Braun

Abstract Characean green algae provide two well-established model cell types for gravitropic research. Experiments in the almost stimulus-free microgravity environments of parabolic flights, sounding rocket flights and Space Shuttle missions have contributed greatly to the progress that has been made in the understanding of cellular and molecular mechanisms underlying plant gravity sensing and gravity-oriented growth responses. While in higher-plant statocytes the role of actin in gravity sensing is still enigmatic, there is clear evidence that actin is intimately involved in polarized growth, gravity sensing and the positive and negative gravitropic response of characean rhizoids and protonemata. The apical tip-growth organizing structure, the Spitzenkörper, and a steep gradient of cytoplasmic free calcium are crucial components of a feedback mechanism that controls polarized growth. Microgravity experiments provided evidence that actomyosin plays a key role in gravity sensing by coordinating the position of statoliths, and, upon gravistimulation, directs sedimenting statoliths to specific gravisensitive areas of the plasma membrane, where they initiate the short gravitropic signalling pathways. In rhizoids, statolith sedimentation is followed by a local reduction of cytoplasmic free calcium resulting in differential growth of the opposite subapical cell flanks—the downward bending. The negative gravitropic response of protonemata is initiated by statolith sedimentation in the apical dome causing actomyosin-mediated relocation of the calcium gradient and displacement of the center of maximal growth towards the upper flank.

Keywords Actin · *Chara* · Gravitropism · Rhizoid · Protonema · Statolith · Tip growth

4.1 Introduction

Charophytes are ubiquitous, filamentous green algae displaying a higher-plant like stem- and leaf-like structure. The main axis is differentiated into nodes and internodes. For characean green algae, gravitropically tip-growing rhizoids and protonemata originating from nodal cells are critical for survival in a rough and constantly changing environment of lakes and rivers. Positively gravitropic—downward growing—characean rhizoids have a root-like function (Braun and Limbach 2006). When pieces of the thallus are ripped off by the water streaming, rhizoids grow out from nodal cells and anchor the thallus segments in the sediment (Fig. 4.1). Characean protonemata are morphologically almost identical but respond negatively gravitropically (Hodick 1993); they develop and grow upward only in darkness (in the absence of blue light) e.g. when the green thallus got buried in the sediment (Fig. 4.1). As soon as they grow out of the soil back into the light, these cells terminate tip growth, divide into nodes and internodes restoring the complexly organized green alga thallus (Braun and Wasteneys 1998a; see Fig. 4.1).

In gravitropism research, unicellular systems like the tip-growing and graviresponding characeen rhizoids and protonemata have been intensively used to study cellular and molecular mechanisms of gravity perception, signalling pathways and the gravitropic responses (Sievers et al. 1996; Braun 1997; Braun and Wasteneys 1998b; Kiss 2000; Braun and Limbach 2006). The tube-like cells with diameters of up to 30 μm are produced by nodal cells of the green thallus and rapidly expand into the surrounding medium by tip growth (Fig. 4.2). Gravity is the most reliable environmental cue and the only one both cell types use for orientation. Although rhizoids and protonemata morphologically look the same they respond oppositely to gravitropic stimulation (Fig. 4.3). In both cell types, the complete gravitropic perception, transduction and response pathways are very short and limited to the apical region of a single cell. That makes them more easily accessible for a variety of experimental approaches than other gravity-sensing cells, e.g. the

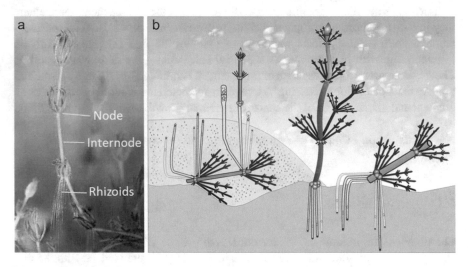

Fig. 4.1 Rhizoids of characean algae originate from nodes of the higher-plant like green thallus (**a**). Rhizoids grow in the direction of gravity (positive gravitropism—(**b**) on the right) to anchor the thallus in the sediment. Protonemata are produced in the absence of blue light (e.g. when the thallus was buried in the sediment) and grow upward against the direction of gravity (negative gravitropism—(**b**) on the left) back into light where they terminate tip growth, divide and regenerate the green thallus. Part of the images were modified after Braun and Limbach (2006)

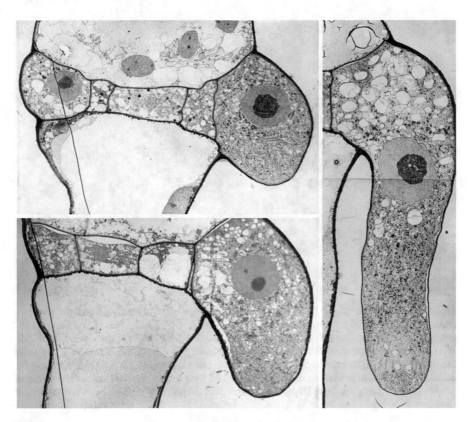

Fig. 4.2 Electron microscopic images showing nodal cells and different stages of outgrowing rhizoids which orientate into the direction of gravity with the onset of polarized growth. Diameter of the rhizoid is 30 μm

statocytes in roots and shoots, in higher plants. Both gravitropic pathways are initiated by a microscopically easy to observe gravity-mediated sedimentation of statoliths, 1–2 μm-sized vacuoles filled with $BaSO_4$ crystals (for review see Sievers et al. 1996; Braun 1997). However, how and where statoliths sediment in the apical dome upon gravitropic stimulation and the subsequent gravitropic response mechanisms are very different in both cell types (Braun 2002).

This chapter summarizes the results that have been collected over the last decades by studying gravity sensing and graviorientation in characean rhizoids and protonemata by means of various molecular, physiological and immunological methods, with innovative advanced microscopic technologies and laser-optical micromanipulation, but most importantly, also by altering gravity conditions by using centrifugation and microgravity-simulation facilities as well as by doing experiments in the almost stimulus-free microgravity environment of spaceflight missions like parabolic flights, sounding rocket flights and Space Shuttle missions. Especially the microgravity experiments provided fascinating new insights and breakthroughs in our understanding of plant gravity sensing and of the processes that underlie the opposite

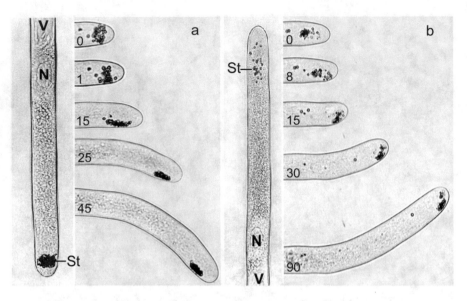

Fig. 4.3 Time series of the positive gravitropic response of a rhizoid (**a**) and the negative gravitropic response of a protonema (**b**) of *Chara globularis*. Upon horizontal positioning (90° gravistimulation), statoliths in both cell types sediment onto specific areas of the lower cell flank. In rhizoids, statolith sedimentation initiates gravitropic bending by differential growth of the opposite cell flanks. The negative gravitropic response of the protonema is initiated by statoliths invading the apical dome and settling close to the cell tip, followed by a drastic relocalization of the center of growth towards the upper flank. Diameter of the cells is 30 μm. Part of the images were modified after Braun (2001)

gravitropic responses in the form of the reorientation of the growth direction in rhizoids and protonemata. We discuss the role of actomyosin forces, actin-binding proteins and calcium as major key players in gravitropic signalling pathways. Since characean algae are thought to be the algae most closely related to land plants (Lewis and McCourt 2004), disclosing the mysteries of gravitropic signalling in rhizoids and protonemata is also a promising approach to provide clues for unravelling the more complex processes of gravity sensing and gravity-oriented growth in higher plants.

4.2   The Cytoskeletal Basis of Gravitropic Tip Growth

Tip growth is an extreme form of polarized growth, one of the most widespread phenomena found in algae, mosses, fungi and higher plant pollen tubes and root hairs; it is also found in tracheids, mesenchyma and endodermal cells. This mode of cell expansion exclusively at the very tip is the preferred growth mode when penetration of the surrounding medium or tissue is required (for review see Heath 1990). Many tip-growing cell types like pollen tubes respond to chemical clues for orienting their growth direction (Hepler et al. 2001). Gravitropic tip growth, however, is limited to a relative small number of cell types like the rhizoids and protonema of characean algae and the apical protonema and caulonema cells of mosses and ferns. In these cell types, the growth direction is strictly governed by the gravity vector (Sievers et al. 1996; Braun 1997; Braun and Limbach 2006).

Characean rhizoids and protonemata exhibit a prominent polar cytoplasmic organization. While actin is the essential cytoskeletal component for all motile processes and is found in all regions of the cell, microtubules maintain the cytoplasmic zonation, stabilize the basal vacuolar zone and the position of the slowly rotating nucleus as well as the organelle distribution in the relatively stationary subapical cytoplasmic region (Fig. 4.4). Microtubules, however, were not found in the apical region and, thus, are not involved in the primary steps of gravitropic signalling and the gravitropic response (Braun and Sievers 1994; Braun and Wasteneys 1998b).

A characteristic feature of tip-growing cells is a vesicle-rich region in the apex. The accumulation of secretory vesicles is part of an apical tip-growth organizing structure called 'Spitzenkörper' or 'apical body' or 'apical clear zone'. Most other

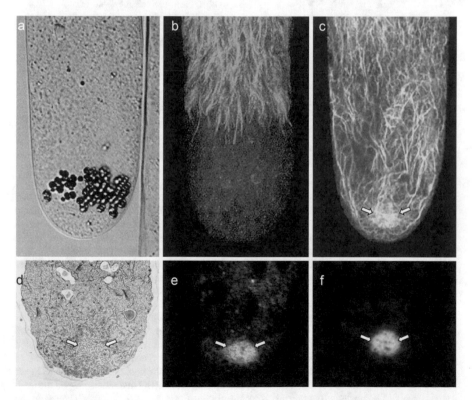

Fig. 4.4 Micrographs of the apical region of characean rhizoids showing a differential interference contrast image (**a**), anti-tubulin immunofluorescence (**b**), rhodamine-phalloidin labelling of the actin microfilament organization (**c**) as well as an electron microscopic image (**d**), anti-ADF (actin-depolymerizing factor) labelling (**e**) and anti-spectrin labelling (**f**). Arrows indicate the position of the center of the Spitzenkörper. Diameter of the rhizoid is 30 μm

organelles like dictyosomes, mitochondria and in most cases also endoplasmic reticulum is excluded from this apical area (Geitmann and Emons 2000; Hepler et al. 2001; Lovy-Wheeler et al. 2005). In contrast to other tip-growing cell types, the Spitzenkörper in characean rhizoids and protonemata is characterized by prominent spherical accumulation of endoplasmic reticulum cisternae that constitutes the structural center of the tip-growth machinery which is surrounded by an accumulation of secretory vesicles of different size and contrast (Figs. 4.4 and 4.5). The

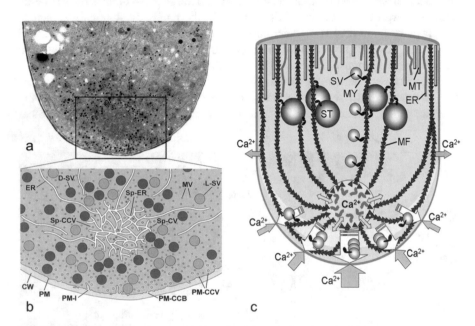

Fig. 4.5 Ultrathin-section electron micrograph of the apical region of a high-pressure frozen and freeze-substituted *Chara* rhizoid (**a**). A dense aggregation of ER cisternae in the center of the growth organizing Spitzenkörper (Sp-ER) is surrounded by an accumulation of secretory vesicles (SV). (**b**) Schematic tomographic model of the outermost apical region of a high-pressure frozen and freeze-substituted rhizoid (position of the modelled region is indicated by the box in (**a**)). Two different types of secretory vesicles (SV), modelled in blue and in light blue, and microvesicles (MV) are evenly distributed in the apical cytoplasm. Clathrin-coated vesicles (CCV) are confined to an approx. 500-nm broad region along the apical plasma membrane (PM). The apical plasma membrane exhibits tube-like intrusions into the cytoplasm as well as extrusions into the cell wall. Bar: 500 nm. (**c**) Schematic illustration of the apical region of *Chara* rhizoid. Actin microfilaments (MFs) with opposite polarities originate from the center of the Spitzenkörper (yellow circle) providing tracks for the myosin (MY)-driven acropetal transport of secretory vesicles (SV) which accumulate in the tip and incorporate cell wall material along the steep tip-high gradient of cytoplasmic free Ca^{2+} (yellow semi lunar area). In normal vertical orientation, statoliths (ST) are kept in a dynamically stable position at a certain distance from the tip by myosins acting on the actin mirofilaments to compensate the apically directed gravity force. The ER cisternae (green) in the center of the Spitzenkörper might function as a storage compartment for calcium and might help to regulate the steepness of the calcium gradient. White arrows indicate the exocytosis rate; grey arrows indicate calcium fluxes at the apical plasma membrane. MT microtubule; ER endoplasmic reticulum. Images were modified after Braun and Limbach (2006) and Limbach and Braun (2008)

position of the Spitzenkörper defines the center of maximal growth, the tip-most plasma membrane area where incorporation of vesicles is maximal (Bartnik and Sievers 1988; Hejnowicz et al. 1977; Sievers et al. 1979; Braun 1996a).

The structural integrity of the Spitzenkörper and its function as the apical tip-growth machinery was found to be strictly dependent on actin, myosin and numerous actin-associated proteins (Braun 1996b, 2001; Braun and Wasteneys 2000; Braun et al. 2004) While information on the arrangement and function of the actin cytoskeleton in the apex of most other tip-growing cell types is scarce, the role of actin in characean rhizoids and protonemata has been well characterized. A complex array of distinct actin microfilaments was demonstrated in the apical dome (Braun and Wasteneys 1998a, b). Fine, mostly axially arranged actin bundles focus in an actin-dense area in the center of the Spitzenkörper in colocalization with the dense spherical aggregate of endoplasmic-reticulum cisternae (Fig. 4.4). Myosin-like motor proteins mediate the transport of secretory vesicles along actin microfilaments that radiate out from the Spitzenkörper towards the apical membrane (Braun 1996b; Braun and Wasteneys 1998b). Such a distinct apical actin array has not been found in any other tip-growing cell types so far.

High-pressure frozen and freeze-substituted rhizoids investigated by dual axis electron tomography (Limbach and Braun 2008) revealed two different types of secretory vesicles as well as microvesicles evenly distributed in the apical region of rhizoids, whereas clathrin-coated vesicles were exclusively found in close vicinity of the apical plasma membrane (Fig. 4.5). The latter vesicles are most likely involved in endocytotic processes mediating the recycling of membrane and the turn-over of membrane-bound proteins such as ion channels and receptor proteins. When the balance of apical exocytotic and endocytotic processes was disturbed by inhibitor-induced disruption of the actin cytoskeleton, the tip-focussed distribution pattern of calcium channels and the steep, tip-high gradient of cytoplasmic free calcium dissipated, and tip growth stopped (Braun and Richter 1999). This steep calcium gradient dictates the incorporation pattern of secretory vesicles and regulates the spatiotemporal activity of actin-binding proteins.

The multiple functions and the dynamic nature of the actin cytoskeleton in rhizoids and protonemata are orchestrated by numerous actin-binding proteins. Spectrin-like proteins, ADF (actin-depolymerizing factor) and profilin were detected in the center of the Spitzenkörper (Fig. 4.4; Braun et al. 2004). Spectrin-like proteins participate in the structural integrity of the ER aggregate by forming crosslinks between ER membranes and actin microfilaments (Braun 2001). Spectrins are known to provide a mechanism for recruiting specific subsets of membrane proteins and to form functional microdomains and, thus, they might help to create the unique physiological conditions for the molecular mechanisms involved in gravity-sensing and polarized growth (Braun 2001).

Furthermore, the presence of ADF and profilin in the center of the Spitzenkörper indicates an actin polymerizing function of this central area (Braun et al. 2004). Additional evidence comes from drug-induced disruption of the actin cytoskeleton by cytochalasin D or the calcium ionophore A23187, which both cause a disintegration of the entire tip-growth organizing structure (Fig. 4.6). After removal of the drugs, resumption of tip-growth is preceded by the recovery of the Spitzenkörper which starts at the apical-most plasma membrane area with the reformation of the ER aggregate and the reorganization of the complex actin arrangement together with the reappearance of the central spectrin and ADF spots (Braun 2001; Braun et al. 2004). These results let us to conclude that the center of the Spitzenkörper functions as an apical actin polymerization site that has not been found in any other tip-growing cell type. We assume that the complexly coordinated, highly dynamic actin architecture in the rapidly extending tip is functionally related to the fundamental role of the actomyosin system in the different phases of gravity sensing and gravity-oriented tip-growth.

The actin-crosslinking protein fimbrin has been localized only in the apical and subapical region suggesting that it is involved in the formation of the mainly axially oriented, dense actin meshwork (Braun et al. 2004). Immunolocalization of the actin-bundling protein villin is restricted to the basal zone of rhizoids and protonemata, where actin microfilaments are separated according to their polarity and form two populations of thick interconnected actin bundles which generate the rotational cytoplasmic streaming around the large central vacuole (Braun et al. 2004).

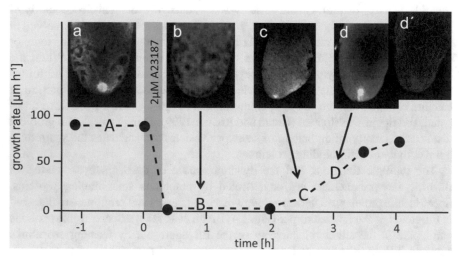

Fig. 4.6 Graph showing the rates of elongation growth of a representative *Chara* rhizoid prior to and after incubation with the calcium ionophore A23187 (2 µM) for 10 min and the corresponding spectrin-immunolabeling images (**a–d**). The loss (**b**) and the gradual reappearance of spectrin-like epitopes in the center of the Spitzenkörper (**c, d**) as well as the reorganization of the actin cytoskeleton (**d´**) is shown during the formation and outgrowth of the new tip after resumption of tip-growth activity. Anti-spectrin immunofluorescence reappears in the form of a small patch at the apical membrane and, later, resumes its original position and size in the center of the Spitzenkörper. Modified after Braun (2001)

4.3 The Positive and Negative Gravisensing Mechanism

As mentioned before, gravity sensing in rhizoids and protonemata is based on gravity-mediated sedimentation of statoliths (Fig. 4.3). Only when rhizoids and protonemata grow exactly in the nominal vertical orientation statoliths are kept in a dynamically balanced position close to the tip. Any deviation of the cell's axis from the direction of gravity is followed immediately by sedimentation of statoliths onto the lateral cell flank, where graviperception occurs and the gravitropic growth response is initiated (Sievers et al. 1996; Braun 2002).

Experiments in microgravity (TEXUS sounding rocket and Space Shuttle missions), on centrifuges and in simulated weightlessness (3D-clinostats, fast-rotating clinostats and random-positioning machines) have provided clear evidence for the complexly organized actomyosin forces both cell types use to precisely balance and regulate the position of the statoliths close to the tip (Sievers et al. 1991a; Volkmann et al. 1991; Braun and Sievers 1993; Buchen et al. 1993, 1997; Hoson et al. 1997; Braun 2002; Krause et al. 2018).

In tip-downward growing rhizoids, the statoliths are located 10–35 μm above the tip, where they do not interfere with the tip growth machinery (Fig. 4.3). The statoliths are kept in this dynamically stable position by two counteracting forces; net-basipetally acting actomyosin forces compensate exactly the gravity force pulling the statoliths into the tip (Fig. 4.7). When the actin microfilaments system is disrupted by inhibitors, statoliths immediately fall freely into the very tip and tip-growth is terminated (Buchen et al. 1993). When the gravity force was reduced to almost zero during TEXUS sounding rocket flights, the statoliths in rhizoids were actively moved basipetally by the remaining actomyosin forces and came to a stop in a new resting position further away from the tip (Fig. 4.7). However, when rhizoids with a disrupted actin cytoskeleton were launched, the statoliths, which had fallen

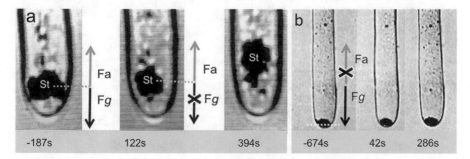

Fig. 4.7 Still-images of the apical region of *Chara* rhizoids videorecorded during the TEXUS 21 (**a**) and TEXUS 25 (**b**) mission. Statoliths, kept in place 10–30 μm above the tip of rhizoids at 1 g before launch, were actively transported away from the tip in the absence of gravity during the 6-min microgravity phase of the rocket flight (**a**). When the actin cytoskeleton was destroyed by cytochalasin D, applied 30 min before launch, statoliths sedimented into the tip of the rhizoid and did not move during the 6-min microgravity phase of the rocket flight (**b**). Modified after Volkmann et al. (1991) and Buchen et al. (1993)

into the tip before launch, did not move at all during the microgravity phase (Fig. 4.7).

Protonemata grow upward, against the direction of gravity, and, accordingly, actomyosin forces act net-acropetally in order to prevent statoliths from sedimenting towards the cell base (Hodick et al. 1998; Braun 2002). Protonemata statoliths are kept in a dynamically stable resting position 10–100 μm below the cell tip (Fig. 4.3).

Microgravity experiments (Buchen et al. 1997; Braun et al. 2002) and optical laser tweezers experiments (Leitz et al. 1995; Braun 2002) have been performed to characterize in detail the complexly arranged actomyosin forces that regulate statolith positioning and transport of statoliths in both rhizoids and protonemata (Fig. 4.8). The laser tweezer force needed to move statoliths towards the apex is much greater than the force required to pull statoliths towards the lateral cell flanks (Braun 2002). During two sounding rocket flights (MAXUS 3 und MAXUS 5)

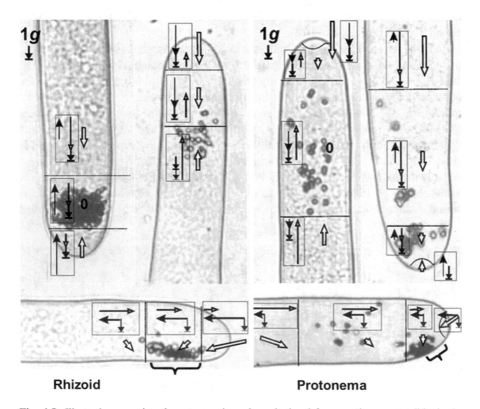

Rhizoid   **Protonema**

Fig. 4.8 Illustration mapping the actomyosin and gravitational forces acting on statoliths in the different apical regions of normal vertically oriented, inverted and horizontally positioned *Chara* rhizoids and protonemata. The gravity force is indicated by truncated arrows, basipetal and acropetal actomyosin forces by arrows with black and white arrowheads, respectively. The resulting force acting on the statoliths is indicated by the white arrows with black outlines. Sedimenting statoliths are directed towards specific gravisensitive areas of the plasma membrane (indicated by brackets). The diameter of the cells is 30 μm. Illustration was redrawn after Braun et al. (2002)

lateral centrifugal forces in a range of 0.1 g were sufficient to trigger a movement of statoliths towards the membrane-bound gravireceptors. From these results, the molecular forces acting on a single statolith in lateral direction were determined to be in a range of 2×10^{-14} N (Limbach et al. 2005), whereas forces required to move statoliths towards the apex were several orders of magnitude higher. All these studies show that by acting differently on statoliths in the different regions of the cells, actomyosin forces ensure that statoliths are actively kept in an area close to the tip, where they can serve as sedimenting particles in the gravitropic signalling pathway. They are free to sediment quickly upon gravistimulation and are close enough to the tip to interfere with the tip-growing process (Braun 2002).

Upon gravistimulation the actomyosin and gravitational forces acting on the statoliths are no longer well balanced and statoliths move towards the lateral cell flanks. Statoliths in both cell types, however, do not simply follow the gravity vector and sediment onto the lateral cell flank; in fact, actomyosin forces direct sedimenting statoliths to specific gravisensitive regions of the plasma membrane (Braun 2002), the only areas, where the gravitropic signalling cascade is elicited triggering the reorientation of the growth direction (Fig. 4.8). Only when rhizoids are reoriented horizontally sedimenting statoliths simply follow the gravity vector and settle onto the lower cell flank 10–35 μm above the tip. However, when rhizoids are gravistimulated at any other angles, sedimenting statoliths do not simply follow the gravity vector, but are (even in inverted rhizoids) actively redirected towards the same belt-like plasma membrane area above the tip where statoliths sediment in horizontally stimulated cells. Centrifugation studies have confirmed that only statolith sedimenting in this specific membrane area were able to trigger graviperception and initiate the positive graviresponse (Braun 2002).

Gravistimulating tip-upward growing protonemata at any angle causes a stimulation-angle independent, actin-mediated acropetal displacement of sedimenting statoliths into the apical dome (Fig. 4.8) where they settle onto the gravisensitive plasma membrane which in protonemata is only a small area very close to the tip, 5–10 μm behind the tip (Braun 2002). Pushing statoliths onto any other plasma membrane area by centrifugation failed to initiate a gravitropic response.

4.4   Gravireceptor Activation Requires Well-Concerted Action of Gravity and Actomyosin Forces

Although the nature of the membrane-bound gravireceptor molecules at the specific areas of the apical plasma membrane in rhizoids and protonemata is still unknown and the immediate downstream physiological processes still need to be clarified, the process of graviperception, the transformation of the physical stimulus of statolith sedimentation into a physiological response, is well understood. There is good experimental evidence from centrifugation, laser-tweezer experiments and parabolic flight experiments that statoliths have to be in contact with the cell-type specific gravisensitive membrane areas in both cell types in order to trigger graviperception and to induce the gravitropic signalling cascade (Braun 2002; Limbach et al. 2005). Lateral movements of statoliths alone, which do not eventually lead to a contact with the plasma membrane, fail to induce a curvature response.

Experiments performed during parabolic flights on board of the A300 Zero-G aircraft ultimately revealed the mechanism of gravireceptor activation in characean rhizoids (Limbach et al. 2005). Statoliths which were weightless during the 22-s microgravity phases but still in contact with the plasma membrane were able to activate the putative membrane-bound gravireceptors. Thus, it is ruled out that the pressure exerted by the weight of statoliths is required for receptor activation; it is the direct contact with the gravisensitive membrane that initiates the gravitropic signalling pathway.

Furthermore, control experiments on ground have demonstrated that increasing the weight of sedimented statoliths by lateral centrifugation did not enhance or accelerate the gravitropic response (Limbach et al. 2005), but graviperception was terminated within seconds after the contact of statoliths with the plasma membrane was lost by inverting gravistimulated rhizoids for a few seconds. These results provide clear evidence that graviperception in characean rhizoids relies on direct contact allowing yet unknown components on the statoliths´ surface to interact with membrane-bound receptors rather than on pressure or tension exerted by the weight of statoliths (Limbach et al. 2005).

4.5   Calcium and Cytoskeletal Forces Govern the Positive and the Negative Gravitropic Response Mechanisms

In characean rhizoids and protonemata, the actomyosin system that plays a crucial role in statolith sedimentation and in the activation of gravireceptors is also essentially involved in the fundamentally different graviresponse mechanisms in both cell types. The smooth downward curvature response of rhizoids has been described as ´bending by bowing´ (Fig. 4.3) since it is the result of reduced growth rates of the lower apical cell flank (Sievers et al. 1979; Hodick 1994; Braun and Limbach 2006). Only in this belt-like gravisensitive plasma membrane area sedimentation of statoliths was shown to result in a drastically reduced concentration of cytoplasmic-free calcium, most likely caused by the inhibition of calcium channels and followed by a locally limited reduction of exocytosis of cell-wall material at the lower cell flank (Fig. 4.9; Sievers et al. 1979; Braun and Richter 1999; Braun 2002). The position of the Spitzenkörper (indicated by spectrin-labelling in Fig. 4.10) in the graviresponding rhizoid was always found to be fixed in its central position in the apical dome. And accordingly, also the tip-most position of the calcium gradient did not change at all during bending (Fig. 4.10). This led us to conclude that the center of maximal growth at the cell tip is not affected during the positive graviresponse in rhizoids (Braun 2002).

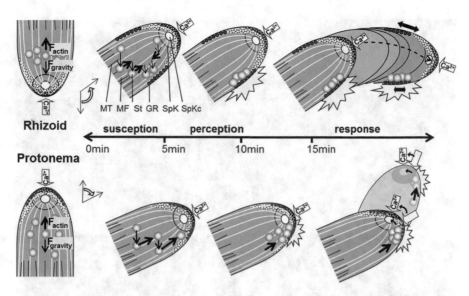

Fig. 4.9 Gravity-sensing mechanisms in characean rhizoids and protonemata. In tip-downward growing rhizoids (upper row), the statolith (St) position results from net-basipetally acting acto-myosin forces (F_{actin}) compensating gravity ($F_{gravity}$). Upon reorientation, statoliths sediment onto the lower cell flank. Net-acropetally acting actomyosin forces in the basal part of the statolith region and in the subapical region prevent statoliths from leaving the apical region and transport the sedimenting statoliths onto membrane-bound gravireceptors (GR) which are restricted to a narrow, beltlike area of the plasma membrane 10–35 μm above the tip. The Spitzenkörper (Spk) remains firmly arrested in the outermost tip and the calcium gradient (indicated by darker and lighter grey dotted area) is always highest at the tip. Statolith sedimentation initiates bending by causing a local reduction of cytosolic Ca^{2+} that results in differential extension of the opposite cell flanks (double-headed arrows). In upward growing protonemata (lower row), the effect of gravity on statoliths is compensated by net-acropetally acting actomyosin forces mediated. Upon horizontal positioning, statoliths settle onto the gravireceptors (GR) which are located near the growth center at the tip by gravity-induced and acropetally directed actomyosin-mediated movements. This causes a drastic shift of the calcium gradient and then of the Spitzenkörper towards the upper flank and the new outgrowth occurs at that site. White arrows point to the area of maximal calcium influx. MT, microtubule; SpKc, center of the Spitzenkörper. Modified after Braun and Limbach (2006)

The negative gravìresponse in protonema has been described as ʹbending by bulgingʹ (Hodick 1994) which refers to the bulge that appears on the upper cell flank at the beginning of the gravìresponse indicating a drastic relocation of the center of maximal growth towards to upper flank (Fig. 4.3). And indeed, the Spitzenkörper (indicated by spectrin labelling in Fig. 4.10) was found to be drastically shifted upward together with the calcium gradient (Fig. 4.10) several minutes after the beginning of the gravìstimulation when sedimenting statoliths intruded into the apical dome and settled close to the very tip, shortly before the bulge appeared on the upper flank of the apical dome (Fig. 4.9).

Indications that the specific properties of actin and specific myosin isoforms, which were shown to be responsible for the anchorage of the Spitzenkörper in the

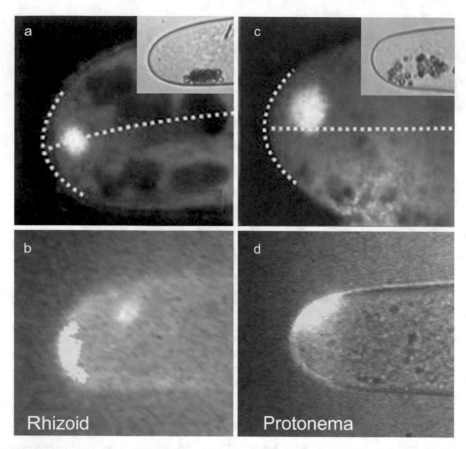

Fig. 4.10 Localization of spectrin epitopes (**a, c**) and the calcium gradient (**b, d**) in a *Chara* rhizoid (insert, left panel) and a protonema (insert, right panel) at the beginning of the graviresponses. The spectrin-spot (**a**) indicating the position of the center of the Spitzenkörper and the calcium gradient (**b**) are still located close to the growth center at the tip at the beginning and during the entire graviresponse in the rhizoids. In the protonemata, the spectrin-spot (**c**) and the calcium gradient (**d**) are clearly shifted towards the upper flank shortly before the outgrowth starts with the formation of a bulge 10 min after gravistimulation. Broken lines outline the outermost tip region and indicate the median line of the cells. The original Calcium Crimson labelling images (**b, d**) are color coded to show the highest pixel intensities in yellow. Diameter of the cells = 30 μm

apical dome, are depending on calcium is strongly supported by fluorescence imaging demonstrating a drastic shift of the steep tip-high calcium gradient and putative calcium channels towards the upper flank shortly before the relocalization of the Spitzenkörper and the initiation of the graviresponse in protonemata (Figs. 4.8 and 4.9). Such a shift of the apical calcium gradient has never been observed in rhizoids (Braun and Richter 1999).

Considering this wealth of data, it is tempting to speculate that the asymmetric influx of calcium mediates the repositioning of the Spitzenkörper and the growth center by differentially regulating the myosin-mediated anchorage and the activity of actin-associated proteins along the shifting calcium gradient (Braun and Richter 1999). Gravistimulation experiments have shown that protonemata tend to reorient towards the former growth axis after only short gravistimulation phases. The position of the newly established growth axis induced by the upward shift of the calcium gradient appears to be rather labile before it is anchored by actin and associated proteins (Braun and Richter 1999; Braun 2001).

4.6   The Impact of Research in Microgravity for Unraveling Plant Gravitropic Signalling Pathways

It all started with a simple experiment. *Chara* rhizoids growing in an agar-filled vacuum-tight chamber on a small microscope payload was put on a TEXUS rocket (TEXUS 21) and launched to an altitude of approx. 360 km. With the beginning of the 6-min microgravity phase of the parabolic flight, statoliths moved away from the tip (Fig. 4.7). For the first time the video-images recorded during the microgravity phase provided direct proof that actomyosin exerts forces on statoliths in a gravity sensing cell (Sievers et al. 1991a; Volkmann et al. 1991; Buchen et al. 1993). When the actin cytoskeleton was destroyed by applying cytochalasin D shortly before launch of another TEXUS rocket, statoliths settled into the very tip of the rhizoids and did not move at áll during the subsequent microgravity phase. It was the first direct proof for the intimate interaction of statoliths with the actin cytoskeleton and unambiguous evidence was obtained for the complex and well-balanced positioning of statoliths in characean rhizoids and protonemata by gravity and actomyosin forces (Buchen et al. 1993). The subsequent studies with rhizoids in microgravity, on clinostats and on centrifuges complemented by laser tweezers studies resulted in the first detailed description of a cytoskeleton-based plant gravity-sensing apparatus (Fig. 4.8; Braun 2002).

In the absence of directional accelerations aboard the Space Shuttle rhizoids developed and grew out from the green thallus nodes in random orientation (Braun 1997). Except for the position of the statoliths, microgravity-developed rhizoids looked identical and showed the same structural polarity as the 1-g controls. This demonstrated that the gravity-sensing cells follow their genetic program and do not require gravity as an environmental cue for development and morphogenesis. Graviresponsiveness was also the same in 1-g and space-grown cells, although the latter had never experienced gravitational forces in their life.

The position of statoliths in microgravity-developed rhizoids (IML-2 Space Shuttle mission) at a greater distance from the cell tip was like that found in rhizoids at the end of the 6-min microgravity phase of TEXUS sounding rocket flights. Although in microgravity the statoliths were positioned further away from the tip, they were still kept in the microtubule-depleted apical zone, where statoliths are able to sediment upon gravistimulation by lateral centrifugation and to initiate the gravitropic response (Braun 1997).

The almost stimulus-free environment of space was also used to study the gravisensitivity of the gravity-sensing mechanisms and to determine gravitropic thresholds in characean rhizoids. Gravisensitivity was first examined in detail during the Spacelab mission IML-2 aboard the Columbia Space Shuttle in 1994. The result of the application of different lateral accelerations was observed by video-microscopy in the slow-rotating centrifuge microscope facility (NIZEMI, Friedrich et al. 1996). The threshold values for gravisensitivity was found to be in the range of 0.1 g and subsequent experiments performed on TEXUS missions and with laser tweezers further narrowed down the minimal molecular force for a lateral displacement (sedimentation) of a statolith towards the gravisensitive plasma membrane area to be in a range of 2×10^{-14} N (Limbach et al. 2005).

And finally, parabolic plane flight experiments were paramount to the very first characterization of a plant gravireceptor activation mechanism. Microgravity experiments provided clear evidence that the membrane-bound gravireceptor molecules in characean rhizoids (of yet unknown nature) are activated upon direct contact with statoliths.

Several reports supporting the idea that characean green algae are the closest relatives of land plants (Turmel et al. 2003; Lewis and McCourt 2004) additionally increase the attractiveness of the characean cell types as model systems for elucidating also the gravitropic sensing mechanism in higher plants. Considering the results from characean unicellular model systems and the evidence for actin-interactions with statoliths in higher plant statocytes obtained from microgravity studies, it is tempting to speculate that actin might play similar roles in the early processes of gravity sensing in higher plants (Volkmann et al. 1991; Hou et al. 2004). However, since disrupting the actin cytoskeleton in statocytes of higher plants does not prevent gravity sensing and the graviresponse at all, actomyosin forces may not be required for the sensory process per se; in fact, the findings suggest that actomyosin may rather have a fine-tuning function by acting as a guiding system and damping modulator for sedimenting statoliths. Thus, actin could ensure an adequate graviresponse by avoiding unfavorable and inappropriate responses to only transient changes in the orientation of the organ with respect to the gravity vector, e.g. when a corn stalk is bending in strong winds.

Nevertheless, there are indications that actomyosin-statolith interactions in higher plant statocytes rely on the same actin-associated proteins and that the mode of gravireceptor activation is very similar relying also on direct contact with higher plant statoliths (unpublished results) rather than on direct or indirect (via actin) mechanical interactions which was postulated by authors since several decades (e.g. see Sievers et al. 1991b; Driss-Ecole et al. 2000; Perbal et al. 2004).

The progress that has been made since the flight of TEXUS 21 in the unravelling of gravitropic signalling pathways further underlines the significance of microgravity research on single cell model systems. Experiments in microgravity and utilization of gravitational and acceleration forces have turned out to be valuable research methods which have contributed greatly to our current knowledge of how plants use the vector information of the gravitational force as a reliable and constant guide for orientation. Space-related experiments including those with centrifuges, clinostats and especially those that have been performed in the almost stimulus-free microgravity environment in parabolic plane flights and sounding rocket flights as well as on Space Shuttle missions have greatly improved our knowledge on the role of the cytoskeleton and associated proteins in the early processes of plant gravity sensing.

The wealth of findings and valuable data from space-related research made gravitropic tip-growing rhizoids and protonemata of characean algae the best understood model cell types for gravity-sensing phenomena in the plant kingdom. The first research project in a microgravity environment heralded an era of breakthroughs in our understanding of the decisive primary phases of gravity sensing in plants.

References

Bartnik E, Sievers A (1988) In-vivo observation of a spherical aggregate of endoplasmic reticulum and of Golgi vesicles in the tip of fast-growing *Chara* rhizoids. Planta 176:1–9

Braun M (1996a) Anomalous gravitropic response of *Chara* rhizoids during enhanced accelerations. Planta 199:443–450

Braun M (1996b) Immunolocalization of myosin in rhizoids of *Chara globularis* Thuill. Protoplasma 191:1–8

Braun M (1997) Gravitropism in tip-growing cells. Planta 203:S11–S19

Braun M (2001) Association of spectrin-like proteins with the actin-organized aggregate of endoplasmic reticulum in the Spitzenkörper of gravitropically tip-growing plant cells. Plant Physiol 125:1611–1620

Braun M (2002) Gravity perception requires statoliths settled on specific plasma-membrane areas in characean rhizoids and protonemata. Protoplasma 219:150–159

Braun M, Limbach C (2006) Rhizoids and protonemata of characean algae – model cells for research on polarized growth and plant gravity sensing. Protoplasma 229:133–142

Braun M, Richter P (1999) Relocalization of the calcium gradient and a dihydropyridine receptor is involved in upward bending by bulging of *Chara* protonemata, but not in downward bending by bowing of *Chara* rhizoids. Planta 209:414–423

Braun M, Sievers A (1993) Centrifugation causes adaptation of microfilaments; studies on the transport of statoliths in gravity sensing *Chara* rhizoids. Protoplasma 174:50–61

Braun M, Sievers A (1994) Role of the microtubule cytoskeleton in gravisensing *Chara* rhizoids. Eur J Cell Biol 63:289–298

Braun M, Wasteneys GO (1998a) Reorganization of the actin and microtubule cytoskeleton throughout blue-light-induced differentiation of characean protonemata into multicellular thalli. Protoplasma 202:38–53

Braun M, Wasteneys GO (1998b) Distribution and dynamics of the cytoskeleton in graviresponding protonemata and rhizoids of characean algae: exclusion of microtubules and a convergence of actin filaments in the apex suggest an actin-mediated gravitropism. Planta 205:39–50

Braun M, Wasteneys GO (2000) Actin in characean rhizoids and protonemata. Tip growth, gravity sensing and photomorphogenesis. In: Staiger CJ, Baluska F, Volkmann D, Barlow PW (eds) Actin: a dynamic framework for multiple plant cell functions. Kluwer, Dordrecht, pp 237–258

Braun M, Buchen B, Sievers A (2002) Actomyosin-mediated statolith positioning in gravisensing plant cells studied in microgravity. J Plant Growth Regul 21:137–145

Braun M, Hauslage J, Czogalla A, Limbach C (2004) Tip-localized actin polymerization and remodeling, reflected by the localization of ADF, profilin and villin, are fundamental for gravity-sensing and polarized growth of characean rhizoids. Planta 219:379–388

Buchen B, Braun M, Hejnowicz Z, Sievers A (1993) Statoliths pull on microfilaments. Experiments under microgravity. Protoplasma 172:38–42

Buchen B, Braun M, Sievers A (1997) Statoliths, cytoskeletal elements and cytoplasmic streaming of *Chara* rhizoids under reduced gravity during TEXUS flights. In: Life sciences experiments performed on sounding rockets (1985–1994), vol 1206. ESA Publications Division, ESA-SP, Nordwijk, pp 71–75

Driss-Ecole D, Jeune B, Prouteau M, Julianus P, Perbal G (2000) Lentil root statoliths reach a stable state in microgravity. Planta 211:396–405

Friedrich ULD, Joop O, Pütz C, Willich G (1996) The slow rotating centrifuge NIZEMI – a versatile instrument for terrestrial hypergravity and space microgravity research in biology and materials science. Biotechnology 47:225–238

Geitmann A, Emons AM (2000) The cytoskeleton in plant and fungal cell tip growth. J Microsc 198:218–245

Heath IB (1990) The role of actin in tip growth of fungi. Int Rev Cytol 123:95–127

Hejnowicz Z, Heinemann B, Sievers A (1977) Tip growth: pattern of growth rate and stress in the *Chara* rhizoid. Z Pflanzenphysiol 81:409–424

Hepler PK, Vidali L, Cheung AY (2001) Polarized cell growth in higher plants. Annu Rev Cell Dev Biol 17:159–187

Hodick D (1993) The protonema of *Chara fragilis* Desv.: regenerative formation, photomorphogenesis, and gravitropism. Bot Acta 106:388–393

Hodick D (1994) Negative gravitropism in *Chara* protonemata: a model integrating the opposite gravitropic responses of protonemata and rhizoids. Planta 195:43–49

Hodick D, Buchen B, Sievers A (1998) Statolith positioning by microfilaments in *Chara* rhizoids and protonemata. Adv Space Res 21:1183–1189

Hoson T, Kamisaka S, Masuda Y, Yamashita M, Buchen B (1997) Evaluation of the three-dimensional clinostat as a simulator of weightlessness. Planta 203:S187–S197

Hou G, Kramer VL, Wang Y-S, Chen R, Perbal G, Gilroy S, Blancaflor EB (2004) The promotion of gravitropism in *Arabidopsis* roots upon actin disruption is coupled with the extended alkalinization of the columella cytoplasm and a persistent lateral auxin gradient. Plant J 39:113–125

Kiss JZ (2000) Mechanisms of the early phases of plant gravitropism. Crit Rev Plant Sci 19:551–573

Krause L, Braun M, Hauslage J, Hemmersbach R (2018) Analysis of statoliths displacement in *Chara* rhizoids for validating the microgravity-simulation quality of clinorotation modes. Microgravity Sci Technol. https://doi.org/10.1007/s12217-017-9580-7

Leitz G, Schnepf E, Greulich KO (1995) Micromanipulation of statoliths in gravity-sensing *Chara* rhizoids by optical tweezers. Planta 197:278–288

Lewis LA, McCourt RM (2004) Green algae and the origin of land plants. Am J Bot 91:1535–1556

Limbach C, Braun M (2008) Electron tomographic characterization of a vacuolar reticulum and of six vesicle types that occupy different cytoplasmic domains in the apex of tip-growing *Chara* rhizoids. Planta 227:1101–1114

Limbach C, Hauslage J, Schaefer C, Braun M (2005) How to activate a plant gravireceptor – early mechanisms of gravity sensing studied in characean rhizoids during parabolic flights. Plant Physiol 139:1–11

Lovy-Wheeler A, Wilsen KL, Baskin TI, Hepler PK (2005) Enhanced fixation reveals the apical cortical fringe of actin filaments as a consistent feature of the pollen tube. Planta 221:95–104

Perbal G, Lefrance A, Jeune B, Driss-Ecole D (2004) Mechanotransduction in root gravity sensing cells. Physiol Plant 120:303–311

Sievers A, Heinemann B, Rodriguez-Garcia MI (1979) Nachweis des subapikalen differentiellen Flankenwachstums im *Chara*-Rhizoid während der Graviresponse. Z Pflanzenphysiol 91:435–442

Sievers A, Kramer-Fischer M, Braun M, Buchen B (1991a) The polar organization of the growing *Chara* rhizoid and the transport of statoliths are actin-dependent. Bot Acta 104:103–109

Sievers A, Buchen B, Volkmann D, Hejnowicz Z (1991b) Role of the cytoskeleton in gravity perception. In: Lloyd CW (ed) The cytoskeletal basis for plant growth and form. Academic Press, London, pp 169–182

Sievers A, Buchen B, Hodick D (1996) Gravity sensing in tip-growing cells. Trends Plant Sci 1:273–279

Turmel M, Otis C, Lemieux C (2003) The mitochondrial genome of *Chara vulgaris*: insights into the mitochondrial DNA architecture of the last common ancestor of green algae and land plants. Plant Cell 15:1888–1903

Volkmann D, Buchen B, Hejnowicz Z, Tewinkel M, Sievers A (1991) Oriented movement of statoliths studied in a reduced gravitational field during parabolic flights of rockets. Planta 185:153–161

Chapter 5
Gravitropism in Fungi, Mosses and Ferns

Donat-Peter Häder

Abstract Motile pseudoplasmodia of cellular and acellular slime molds show gravitaxis while unicellular *Dictyostelium* amoebae do not. Fungi display gravitropism of their fruiting bodies which is thought to facilitate spore dispersal. Caulonemata and sporophytes of liverworts and mosses show negative gravitropism which is a also observed in sporophytes of ferns. Several hypotheses have been proposed to explain the mechanism for graviperception but none has been proven yet.

Keywords Slime molds · Fungi · Liverworts · Mosses · Ferns · Gravitaxis · Gravitropism

5.1 Introduction

Terrestrial organisms including fungi, mosses, ferns and higher plants use the gravity vector of the Earth to optimize their position in their environment (Blancaflor and Masson 2003). It is obvious that photosynthetic organisms orient their shoots upwards in order to maximize photosynthesis (negative gravitropism). Leaves or leaflets are often oriented at an oblique angle, which exposes the laminas perpendicular to the incident sun rays (Franklin and Whitelam 2004; Mano et al. 2006). Roots or rhizoids are usually oriented downwards (positive gravitropism or at an angle to the gravity vector to anchor the plant in the substratum (Cove 1992; Aloni et al. 2006). Even though fungi are not photosynthetic organisms, gravitropic orientation is beneficial to bring the fruiting bodies above the soil's surface and to facilitate the spreading of the spores.

As detailed in the introductory chapter of this volume, gravitropism—as well as gravitaxis in motile microorganisms–is based on a gravireceptor, a structure which senses the pressure of a heavy element, which, under the effect of gravity, exerts pressure on the sensitive element. The pressing element can be a statolith—a heavy organic or inorganic particle such as an amyloplast or an organelle containing barium-sulfate crystals (Sack 1997; Braun 2002). Alternatively, the whole cytoplasm of a cell

M. Braun et al., *Gravitational Biology I*, SpringerBriefs in Space Life Sciences, https://doi.org/10.1007/978-3-319-93894-3_5

can exert pressure to an underlying membrane system. In some cases, mechanosensitive ion channels have been identified as gravisensors, while in other systems, the structural and biochemical identity still has to be provided.

Probably all organisms have developed the ability to sense the direction of the gravity vector of the Earth. Most are capable to respond to it with the exception of very small bacteria and viruses, which are subject to Brownian movement. However, the molecular mechanisms of graviperception may have been evolved several times.

5.2   Slime Molds and Fungi

The cellular slime mold *Dictyostelium discoideum* undergoes a life cycle, which includes a unicellular phase that consists of motile amoebae feeding on soil bacteria (Devreotes 1989). Under starved conditions some amoebae send out a cAMP signal, which is relayed through the population and results in a migration towards the signal center (Henderson 1975). There, 10^3–10^5 amoebae aggregate to form a multicellular pseudoplasmodium (also called slug), which is about 0.5–2 mm long and 0.1 mm in diameter (Loomis 2012). The slugs are also motile and respond to a number of external stimuli including chemical clues, temperature gradients, light and gravity (Poff and Häder 1984; Häder and Hansel 1991; Maree et al. 1999; Song et al. 2006). They move for a few days until they culminate and form a sporangiophore with unicellular spores. Under optimal environmental conditions these spores are shed and form new motile amoebae.

When allowed to migrate on a vertical agar slab in a Petri dish the slugs showed a pronounced positive gravitaxis moving downward in contrast to the unicellular amoebae, which moved in random directions (Fig. 5.1). However, when a light stimulus was given simultaneously, gravitaxis was expressed only at low fluence

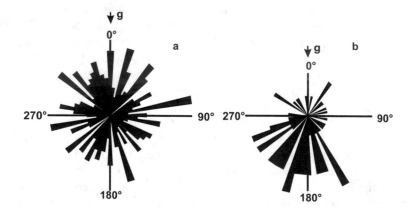

Fig. 5.1 Circular histograms of the distribution and orientation of *Dictyostelium discoideum* amoebae (**a**) and slugs (**b**) on a vertical agar plate. 0° indicates top of the plate. Redrawn after Häder and Hansel (1991)

rates (Häder and Hansel 1991). The pseudoplasmodia are also remarkable since they show positive thermotaxis when exposed to a gradient of 0.2 °C cm^{-1} ignoring the gravistimulus. Irradiances $> 10^{-3}$ W m^{-2} canceled the response with respect to gravity.

In contrast to the cellular slime mold *Dictyostelium*, *Physarum polycephalum* belongs to the acellular slime molds, which form undivided plasmodia with millions of nuclei (Guttes et al. 1961). The plasmodia consist of interconnected strands with a tough ectoplasm and a low-viscosity endoplasm which moves back and forth at regular intervals controlled by a Ca^{2+} oscillator (Coggin and Pazun 1996). When the shuttle streaming occurs more in one direction than in the other the whole plasmodium migrates in that direction. Also these myxomycetes respond to environmental stimuli including light, chemicals and gravity (Sauer 1982; Wolke et al. 1987; de Lacy Costello and Adamatzky 2013). On a vertical surface *Physarum* plasmodia show predominantly positive gravitaxis (68%), while 29% were indifferent and 13% showed negative gravitaxis (Wolke et al. 1987). When the vertical agar plate was rotated stepwise by 90°, the plasmodia reoriented accordingly. This graviresponse was investigated in real microgravity during the IML-1 mission on the Space Shuttle (Block et al. 1992) as well as under simulated microgravity on a fast rotating clinostat (Block et al. 1994). While no obvious statoliths can be detected in the plasmodia, mitochondria have been suggested to exert the pressure on a putative gravireceptor. It is interesting to note that the signal transduction chain involves cAMP which has been demonstrated in graviresponses of other organisms such as ciliates and flagellates as well (Block et al. 1998). Exposure to a few days of microgravity during a space experiment (STS-69) decreased the cAMP concentration in the plasmodia which was, thus, identified as a secondary messenger.

Fungal zoospores show a pronounced negative gravitaxis. The discussion, whether the spores of *Phytophthora* respond to an oxygen gradient, a light signal or a chemical clue, could be solved by proving that they reacted to gravity (Cameron and Carlile 1977). The cells also show chemotaxis, which was shown to be controlled by changes in the membrane potential, which regulates the flagellar activity; however, proof is still lacking, if the same mechanism is responsible for graviorientation (Cameron and Carlile 1980).

Mushrooms show a number of different responses to various environmental stimuli at different times of their development (Moore 1991). They display thigmotropism, gravitropism, anemotropism and phototropism. Young fruiting bodies grow perpendicular away from the substratum even in the presence of increased accelerations. Afterwards, the fruiting body shows positive phototropism and subsequently negative gravitropism; this shift is correlated with the initiation of spore formation. One interpretation for the onset of gravitropism is the assumption that the hymenia (forming to tubes or gills) in the fruiting body should be oriented vertically to facilitate the liberation of the spores falling downward in the free space. In addition, gravity is required for the morphogenesis of the fruiting body (Corrochano and Galland 2006). On an orbiting space station, *Polyporus brumalis* failed to initiate fruiting. Clinostat experiments also indicated that sporulation in *Lentinus tigrinus* and *P. brumalis* is prevented in simulated microgravity and karyogamy was suppressed. When *Coprinus cinereus* was grown on a clinostat, it produced normal fruit body primordials but failed to produce spores. Even though the research on gravitropism in higher fungi (mushrooms) has been carried out for more than 125 years, the underlying mechanism still has to be revealed.

The fruiting bodies of the basidiomycete *Flammulina velutipes* show a clear negative gravitropic response. When oriented horizontally within 2 h the stems of the fruiting body respond with an increased elongation of the lower side, while it decreased by 40% in the upper side (Monzer et al. 1994; Kern et al. 1997). This growth response results in an overshoot, which is subsequently regulated. Under real microgravity in a space experiment, the fruiting bodies showed random growth orientation (Fig. 5.2) (Kern and Hock 1993). The graviresponsiveness seems to be restricted to the apical part of the stipe, which forms the transition zone to the pileus (cap or head). Light and electron microscopy showed that the hyphae in this zone are smaller and less vacuolated than in the basal part of the stipe. Complete removal of the pileus did not affect the gravitropic response, while excision of the transition zone abolished the gravitropic bending. In *Coprinus cinereus* gravitropic bending

Fig. 5.2 Flammulina velutipes. Negative gravitropic orientation of Flammulina velutipes fruiting bodies grown for 5 days on Earth (**1 g**); fruiting bodies grown for 165 h in microgravity in Spacelab during the D2 mission show random orientation (**μg**). Modified after Kern and Hock (1993)

starts even 30 min after the stems have been placed horizontally, however, only after completion of meiosis (Kher et al. 1992).

While several organisms have been found to use heavy statoliths or a heavy cytoplasm to exert pressure on an underlying gravireceptor, an opposite mechanism has been suggested for several fungi. In the zygomycete *Phycomyces blakesleeanus* some 200 lipid globuli are arranged in a spherical complex about 100 μm below the growing tip of the vegetative sporangiophore (Grolig et al. 2004, 2006). The complex is held in place by a dense framework of filamentous actin. The buoyancy of these globuli could exert an upward pressure onto a gravireceptive structure. Experimental inhibition of the globuli formation by growing the sporangiophores at low temperatures reduces the gravitropic response. Similar lipid globuli have been found in the gravitropically responding hyphae of the fungus *Gigaspora margarita* and other fungi. In contrast, Eibel and coworkers suggest that gravitropism is instrumentalized by octahedral protein crystals with a specific mass of 1.2 g cm^{-3} located in the central vacuoles of the sporangiophore acting as statoliths (Eibel et al. 2000). Gravitropic mutants lack these protein crystals. Another publication offered a combined hypothesis based on both buoyant lipid globuli and sedimenting protein crystals. Molecular genetic approaches, magnetophoresis and laser ablation have supported the hypothesis that the actin cytoskeleton is involved in the gravitaxis and gravitropism sensory transduction chain (Kiss 2000).

5.3 Bryophytes

Dark-grown caulonemata and gametophores of the moss *Physcomitrella patens* show a pronounced negative gravitropism (Jenkins et al. 1986). After being placed horizontally, the caulonemata bend about 20° within 12 h and subsequently complete the 90° bending at a slower pace. Several mutants have been found which show a partially or completely inhibited gravitropism; one mutant even shows positive gravitropism. Negative gravitropism in wild-type protonema cells is reversed after a period of growth on a clinostat. The same reversal of the growth direction is observed during mitotic division (Knight and Cove 1991). Protoplast fusion resulting in somatic hybrids showed that at least three genes are involved in gravitropism. It is interesting to note that in none of these mutants gravitropism of the gametophores is affected indicating that the mechanisms of graviperception or transduction are different in caulonemata and gametophores. In contrast to caulonemata, rhizoids show a pronounced positive gravitropism and a negative phototropism (Glime 2017).

Amyloplasts have been discussed as possible statoliths in the protonemata of *Ceratodon purpureus* (Walker and Sack 1990). In the tip, there is a cluster of non-sedimenting amyloplasts with an amyloplasts-free zone below. The amyloplasts below this zone seem to be anchored by (actin?) filaments as they do not sediment to the basal wall, but to the lower cell wall in horizontal protonemata. This behavior resembles that of the barium-sulfate filled statoliths in characean rhizoids. When placed horizontally, wild-type *C. purpureus* protonemata shortly bend downwards, which also occurs prior to cytokinesis (Wagner et al. 1997). UV-induced mutants also show negative gravitropism with kinetics similar to wild-type protonemata. Also *Funaria* caulonemata show upward bending (Schwuchow et al. 1995). The tip cells have a broad subapical zone, where plastid sedimentation has been observed. Under real microgravity on the Space Shuttle mission STS-87 and under simulated microgravity, protonemata cultures showed a radial outgrowth followed by a clockwise spiral growth (Kern et al. 2005). The protonemata of *C. purpureus* also show phototropism in red light. At irradiances ≥ 140 nmol m^{-2} s^{-1} gravitropism was quenched, but amyloplast sedimentation was still observed (Kern and Sack 1999). These results show that both stimuli compete and that light regulates the gravitropic response.

In roots of higher plants, the gravitropic signal is relayed from the statenchyma in the root tip to the elongation zone via the plant hormone auxin, which is guided by several auxin transporters (cf. Chap. 6). Auxins and cytokinins have also been found in mosses and liverworts, where they regulate morphological development (Sabovljević et al. 2014). In *Marchantia polymorpha,* auxin is involved in the establishment of a dorsiventral polarity. Gemmae cups usually respond negative gravitropically, but after external application of auxin, they became positive gravitropic (Flores-Sandoval et al. 2015). These results indicate that the auxin regulation of growth and morphological development has been established in these earliest land plants.

5.4   Ferns

Trophophylls (sterile leafs) of the eusporangiate fern *Danaea wendlandii* show negative gravitropism, while sphorophylls (fertile leafs) grow horizontally (Sharpe and Jernstedt 1990). This behavior has been attributed to a statolith mechanism since sedimenting amyloplasts have been detected in the cells of the petiole and rachis. Also dark-grown gametophytes of *Ceratopteris richardii* show negative gravitropism while light inhibits gravitropism (Kamachi and Noguchi 2012). Also in most *Selaginella* species light modulates the gravitropic response of rhizoids during the early developmental stage (Liu and Sun 1994). Later the effect of light is reduced while gravitropism dominates. Adult plants of the fern *Ceratopteris richardii* show pronounced graviperception and gravitropism. Single-celled spores of this fern were exposed to microgravity during a shuttle space flight (STS-93). cDNA microarray and Q RT-PCR analysis of spores germinating in microgravity showed significant changes in the mRNA expression of about 5% of the analyzed genes (Salmi and Roux 2008). Similar changes in gene expression are found in animal and plant cells.

References

Aloni R, Aloni E, Langhans M, Ullrich C (2006) Role of cytokinin and auxin in shaping root architecture: regulating vascular differentiation, lateral root initiation, root apical dominance and root gravitropism. Ann Bot 97:883–893

Blancaflor EB, Masson PH (2003) Plant gravitropism. Unraveling the ups and downs of a complex process. Plant Physiol 133:1677–1690

Block I, Wolke A, Briegleb W, Wohlfarth-Bottermann KE, Merbold U, Brinckmann E, Brillouet C (1992) Graviresponse of Physarum – investigations in actual weightlessness. Cell Biol Int Rep 16:1097–1102

Block I, Wolke A, Briegleb W (1994) Gravitational response of the slime mold Physarum. Adv Space Res 14:21–34

Block I, Rabien H, Ivanova K (1998) Involvement of the second messenger cAMP in the gravity-signal transduction in Physarum. Adv Space Res 21:1311–1314

Braun M (2002) Gravity perception requires statoliths settled on specific plasma membrane areas in characean rhizoids and protonemata. Protoplasma 219:150–159

Cameron JN, Carlile MJ (1977) Negative geotaxis of zoospores of the fungus Phytophthora. J Gen Microbiol 98:599–602

Cameron JN, Carlile MJ (1980) Negative chemotaxis of zoospores of the fungus Phytophthora palmivora. J Gen Microbiol 120:347–353

Coggin SJ, Pazun JL (1996) Dynamic complexity in Physarum polycephalum shuttle streaming. Protoplasma 194:243–249

Corrochano L, Galland P (2006) Photomorphogenesis and gravitropism in fungi. In: Kües U, Fischer R (eds) Growth, differentiation and sexuality. The mycota (A comprehensive treatise on fungi as experimental systems for basic and applied research). Springer, Berlin, pp 233–259

Cove DJ (1992) Regulation of development in the moss, Physcomitrella patens. In: Brody S, Cove DJ (eds) Developmental biology. A molecular genetic approach. Springer, Berlin, pp 179–193

de Lacy Costello B, Adamatzky AI (2013) Assessing the chemotaxis behavior of Physarum polycephalum to a range of simple volatile organic chemicals. Commun Integr Biol 6:e25030

Devreotes P (1989) Dictyostelium discoideum: a model system for cell-cell interactions in development. Science 245:1054–1058

Eibel P, Schimek C, Fries V, Grolig F, Schapat T, Schmidt W, Schneckenburger H, Ootaki T, Galland P (2000) Statoliths in Phycomyces: characterization of octahedral protein crystals. Fungal Genet Biol 29:211–220

Flores-Sandoval E, Eklund DM, Bowman JL (2015) A simple auxin transcriptional response system regulates multiple morphogenetic processes in the liverwort Marchantia polymorpha. PLoS Genet 11:e1005207

Franklin KA, Whitelam GC (2004) Light signals, phytochromes and cross-talk with other environmental cues. J Exp Bot 55:271–276

Glime JM (2017) Ecophysiology of development: protonemata. In: Glime JM (ed) Bryophyte ecology. Michigan Technological University and the International Association of Bryologists, Houghton, MI

Grolig F, Herkenrath H, Pumm T, Gross A, Galland P (2004) Gravity susception by buoyancy: floating lipid globules in sporangiophores of Phycomyces. Planta 218:658–667

Grolig F, Döring M, Galland P (2006) Gravisusception by buoyancy: a mechanism ubiquitous among fungi? Protoplasma 229:117–123

Guttes E, Guttes S, Rusch HP (1961) Morphological observations on growth and differentiation of Physarum polycephalum grown in pure culture. Dev Biol 3:588–614

Häder D-P, Hansel A (1991) Response of Dictyostelium discoideum to multiple environmental stimuli. Bot Acta 104:200–205

Henderson EJ (1975) The cyclic adenosine 3′: 5′-monophosphate receptor of Dictyostelium discoideum. Binding characteristics of aggregation-competent cells and variation of binding levels during the life cycle. J Biol Chem 250:4730–4736

Jenkins GI, Courtice GRM, Cove DJ (1986) Gravitropic responses of wild-type and mutant strains of the moss *Physcomitrella patens*. Plant Cell Environ 9:637–644

Kamachi H, Noguchi M (2012) Negative gravitropism in dark-grown gametophytes of the fern *Ceratopteris richardii*. Am Fern J 102:147–153

Kern VD, Hock B (1993) Gravitropism of fungi – experiments in space. Life sciences research in space, proceedings of the fifth European symposium. European Space Agency, Arcachon

Kern VD, Sack FD (1999) Irradiance-dependent regulation of gravitropism by red light in proto-nemata of the moss *Ceratodon purpureus*. Planta 209:299–307

Kern VD, Mendgen K, Hock B (1997) *Flammulina* as a model system for fungal graviresponses. Planta 203:23–32

Kern VD, Schwuchow JM, Reed DW, Nadeau JA, Lucas J, Skripnikov A, Sack FD (2005) Gravitropic moss cells default to spiral growth on the clinostat and in microgravity during spaceflight. Planta 221:149–157

Kher K, Greening JP, Hatton JP, Frazer LN, Moore D (1992) Kinetics and mechanics of stem gravitropism in *Coprinus cinereus*. Mycol Res 96:817–824

Kiss JZ (2000) Mechanisms of the early phases of plant gravitropism. Crit Rev Plant Sci 19:551–573

Knight CD, Cove DJ (1991) The polarity of gravitropism in the moss *Physcomitrella patens* is reversed during mitosis and after growth on a clinostat. Plant Cell Environ 14:995–1001

Liu B, Sun G (1994) Effect of light on gravitropic response of rhizoids of gametophytes of ferns. Wuhan Bot Res 12:165–169

Loomis W (2012) *Dictyostelium discoideum*: a developmental system. Elsevier, Groningen

Mano E, Horiguchi G, Tsukaya H (2006) Gravitropism in leaves of *Arabidopsis thaliana* (L.) Heynh. Plant Cell Physiol 47:217–223

Maree AFM, Panfilov AV, Hogeweg P (1999) Migration and thermotaxis of *Dictyostelium discoideum* slugs, a model study. J Theor Biol 199:297–309

Monzer J, Haindl E, Kern V, Dressel K (1994) Gravitropism of the basidiomycete *Flammulina velutipes*: morphological and physiological aspects of the graviresponse. Exp Mycol 18:7–19

Moore D (1991) Perception and response to gravity in higher fungi—a critical appraisal. New Phytol 117:3–23

Poff KL, Häder D-P (1984) An action spectrum for phototaxis by pseudoplasmodia of *Dictyostelium discoideum*. Photochem Photobiol 39:433–436

Sabovljević M, Vujičić M, Sabovljević A (2014) Plant growth regulators in bryophytes. Bot Serbica 38:99–107

Sack FD (1997) Plastids and gravitropic sensing. Planta 203:63–68

Salmi ML, Roux SJ (2008) Gene expression changes induced by space flight in single-cells of the fern *Ceratopteris richardii*. Planta 229:151–159

Sauer HW (1982) Developmental biology of *Physarum*. Cambridge University Press, Cambridge

Schwuchow JM, Kim D, Sack FD (1995) Caulonemal gravitropism and amyloplast sedimentation in the moss *Funaria*. Can J Bot 73:1029–1035

Sharpe JM, Jernstedt JA (1990) Tropic responses controlling leaf orientation in the fern *Danaea wendlandii* (Marattiaceae). Am J Bot 77(8):1050–1059

Song L, Nadkarni SM, Bödeker HU, Beta C, Bae A, Franck C, Rappel W-J, Loomis WF, Bodenschatz E (2006) *Dictyostelium discoideum* chemotaxis: threshold for directed motion. Eur J Cell Biol 85:981–989

Wagner TA, Cove DJ, Sack FD (1997) A positively gravitropic mutant mirrors the wild-type protonemal response in the moss *Ceratodon purpureus*. Planta 202:149–154

Walker LM, Sack FD (1990) Amyloplasts as possible statoliths in gravitropic protonemata of the moss *Ceratodon purpureus*. Planta 181:71–77

Wolke A, Niemeyer F, Achenbach F (1987) Geotactic behavior of the acellular myxomycete *Physarum polycephalum*. Cell Biol Int Rep 11:525–528

Chapter 6
Gravitropism in Higher Plants: Cellular Aspects

Dennis Said Gadalla, Markus Braun, and Maik Böhmer

Abstract Due to their sessile life style, an important ability of plants is to adjust their growth towards or away from environmental stimuli. Plant responses that involve directed movements are called tropisms. Among the best-known tropisms are phototropism, the response to light, and gravitropism, the response to gravity. Gravity is one of the major factors that govern root growth in plants. Since the emergence of land plants, gravitropism allowed plants to adjust root growth to maximize access to water and nutrients, and shoots to explore and exploit space on and above the surface of the Earth. In this chapter we discuss current knowledge and point out open questions like the nature of the gravireceptor, the role of secondary messengers, hormones and the cytoskeleton. We review the history of plant gravitropism research, from early experiments performed by naturalists like Charles Darwin to the utilization of clinostats, centrifuges and experimentation in the almost stimulus-free environment of microgravity provided by drop towers, parabolic flights of aircrafts and rockets, satellites and low earth orbit space stations, which are increasingly contributing to our understanding of plant gravity sensing and orientation.

Keywords Auxin · Clinostat · Gravitropism · Gravity · Microgravity · Roots · Statolith

6.1 Introduction

All organisms on Earth are subject to the continuous influence of gravity. In a process called gravitropism, plants perceive the direction of gravity and can adjust their growth accordingly (Fig 6.1). Most plant organs are actively positioned at various defined angles from the gravity vector, the gravitational set point angle (GSA; Digby and Firn 1995).

This allowed plants to leave the water and conquer land, enabling them to explore and exploit space below, on and above the surface of the Earth in a most beneficial way and, thus, to provide food and resources for all animal and human life on

M. Braun et al., *Gravitational Biology I*, SpringerBriefs in Space Life Sciences,
https://doi.org/10.1007/978-3-319-93894-3_6

Fig. 6.1 Higher plant organs make use of the only constant environmental cue, the gravity vector of the Earth, for precisely orienting their organs. By orienting positive gravitropic roots they grow towards the center of gravity to anchor themselves in the soil and take up water and nutrients. Shoots grow negatively gravitropically away from the center of the Earth towards light. Gravitropism allows plants to arrange their different organs in a most beneficial way to exploit the space below, on and above the surface of the Earth

Earth. An organ like the primary root that is maintained vertically and grows downwards towards the gravity vector has a GSA of 0°, an extreme called positive gravitropism. An organ like the shoot that is maintained vertically but grows upwards against the gravity vector has a GSA of 180°, an extreme termed negative gravitropism (Frank 1868). Other organs do not show strict positive or negative gravitropism but have a GSA between these two extremes (plagiotropism;

Shuttle gardening   USA TODAY · MONDAY, NOVEMBER 4, 1985

Fig. 6.2 Dutch astronaut Wubbo Ockels displays garden cress sprouts growing inside Spacelab aboard the space shuttle Challenger

cf. Fig. 1.2). One example is the gravitropic behavior of lateral roots in *Arabidopsis thaliana*. These bud from the main root orthogonal to the gravitational vector (diatropism) and upon reorientation to either a more upright or downward facing direction, the tip of the lateral root will display a rapid curvature that brings it back towards the original orientation (Mullen and Hangarter 2003; Guyomarc'h et al. 2012).

Research on gravity sensing and gravitropic responses has greatly benefitted from the advancement in molecular and cellular methods as well as from experiment platforms providing the almost stimulus-free environment of microgravity. During the German Spacelab-mission D1, US newspapers published a photograph of the Dutch astronaut Wubbo Ockels showing roots of the garden cress, which germinated and grew on board the Space Shuttle Challenger (see Fig. 6.2). In the absence of gravity, the roots grew straight in the direction given by the tip of the radicle in the seed, thus, confirming that Wilhelm Pfeffer's predicted "automorphose" really exists at the organ and the cellular level (Pfeffer 1904). It was found that the architecture and function of gravity perceiving cells, the statocytes, are genetically determined and not dependent on the presence of gravity or other factors (Volkmann et al. 1986). Since then, numerous microgravity experiments performed in drop towers, on parabolic plane flights, on research rockets, on Russian and Chinese satellites and on US Space shuttles as well as on the International Space Station have opened new perspectives on the molecular, cellular and physiological mechanisms underlying gravity sensing and graviorientation in higher plants.

6.2   Gravity Perception

The gravitropic response can be subdivided into three distinct and sequential events: signal perception, signal transduction and gravitropic response. The first step in signal perception is the conversion of the gravitational stimulus into a detectable physical change within the cell (Baldwin et al. 2013; Toyota and Gilroy 2013; Schüler et al. 2015). Various models have been proposed of how the physical stimulus of statolith sedimentation is transduced into a biochemical signaling event. The leading hypotheses are the statolith-dependent starch-statolith hypothesis and the tensegrity model (Yoder 2001; Zheng and Staehelin 2001) as well as the statolith-independent gravitational pressure or protoplast pressure models (Wayne and Staves 1996; Palmieri and Kiss 2007).

6.2.1   Tissue Localization of Graviperception

All plant cells experience the gravitational stimulus equally. However, one must differentiate between a specific gravity sensing mechanism that is localized in specific tissues, and general unspecific physical and physiological or metabolic reactions that might be observable in all cells. For a gravitropic response, gravity must be perceived by specialized tissues (Fig. 6.3). Graviperception in the shoot occurs mainly in the starch sheath in the endodermis. In *Arabidopsis thaliana*, a mutation of the gene *PHOSPHOGLUCOMUTASE* (*PGM*) prevents formation of the starch sheath, phenotypically observable by an agravitropic behavior of the shoot. This connection between defect and absent endodermis and gravitropism was further confirmed by mutant screens. The mutant *shoot gravitropism 1* (*sgr1*) has a mutation in the gene *SCARECROW* (*SCR*), the mutant *sgr7* has a mutation in gene *SHORT ROOT* (*SHR*). In both mutants, the endodermis is not formed in the shoot and consequently, in both mutants shoot gravitropism is no longer observable (Fukaki et al. 1998). All *shoot gravitropism* mutants, although likewise defective in root endodermis, still show normal root gravitropism which is easily explained because in the root, gravity is sensed in the columella cells in the root cap.

Charles Darwin was one of the first naturalists and biologists who had shown that the removal or damaging of the root tip leads to agravitropic growth and that these roots regain the ability to grow downward again after regeneration of the root cap (Darwin 1880). In particular the manipulation or removal of the columella cells impairs gravitropic behavior (Juniper et al. 1966; Konings 1968; Tsugeki and Fedoroff 1999). There are typically 4 layers of columella cells at the root tip, S1-S4 (Fig. 6.3). By selective removal of individual columella cells using laser ablation, it was shown that the innermost layers S1 and S2 are most important for gravitropism (Blancaflor et al. 1998).

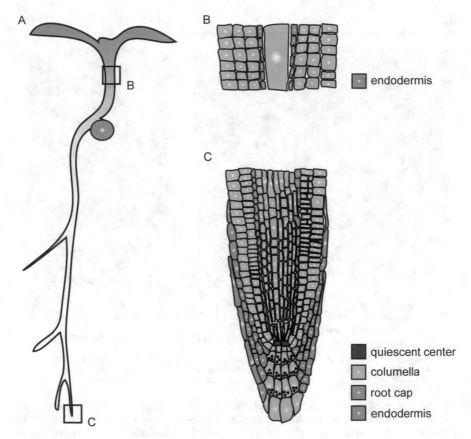

Fig. 6.3 Tissues of graviperception. The endodermis in the shoot and the columella cells in the root tip form gravity sensing statocytes

6.2.2   Starch-Statolith Hypothesis

The starch-statolith hypothesis for plant gravity sensing is widely accepted by most researchers today. Statocytes are highly specialized gravity-sensing cells of higher plants and are characterized by a polar architecture (first described in detail by Sievers and Volkmann 1971, 1977) and the presence of starch-filled amyloplasts, the so-called statoliths (Figs. 6.3 and 6.4). Statocytes in roots have small vacuoles, a nucleus that is positioned in the upper part of the cell by cytoskeletal elements, and a specific architecture of the endoplasmic reticulum (ER). The ER in statocytes is organized in two forms. Most of the ER is tubular and covers the periphery of the cell, the so-called cortical ER. In the lower part of most statocytes, ER cisternae form a cushion-like structure onto which statoliths seem to be already sedimented in statocytes in a normal vertically growing root. The starch-filled amyloplast-statoliths are the only organelles free to move and to sediment in the cell upon gravistimulation,

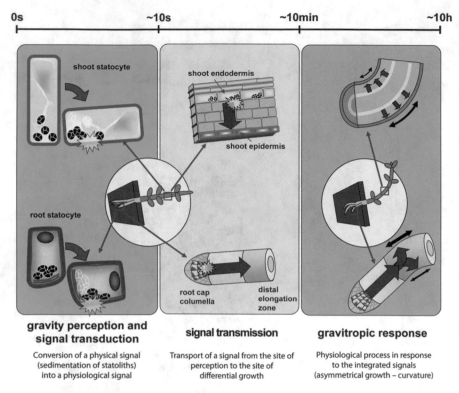

Fig. 6.4 Phases of gravitropism in higher plants. When the orientation of a plant organ changes relative to the vector of gravity, sedimentation of amyloplast statoliths in statocytes initiates signal perception—the conversion of a physical signal of statolith sedimentation into a physiological signal. This signal is then transmitted to the responding target cells. The nominal orientation is resumed by differential growth of the opposite flanks of the organs

probably favored by the absence of prominent actin microfilament bundles, extensive microtubule arrangements (Baluška et al. 1997) and ER cisternae in the central area as well as the proximal position of the nucleus and the starch content of the amyloplast-statoliths. All other organelles in statocytes as well as the organelles in all other cell types are precisely fixed in their position by cytoskeletal elements.

Upon reorientation statoliths sediment along the gravitational vector. Already at the beginning of the twentieth century, this was considered a trigger for gravitropic responses (Haberlandt 1900; Nemec 1900). Convincing evidence in favor of the statoliths-based sensing mechanism comes from magnetophoretic experiments with barley coleoptiles and flax roots as well as in moss protonemata (*Ceratodon purpureus*) showing clearly that a lateral displacement of statoliths without tilting the plant organs or the protonemata from the vertical growth direction is sufficient to trigger typical gravitropic responses independent of the gravitational vector (Kuznetsov and Hasenstein 1996, 1997; Kuznetsov et al. 1999). Cold treatment, starvation and destarching amyloplast-statoliths with gibberellic acid or kinetin reversibly reduced or even completely abolished graviresponsiveness in cress roots (Audus 1979; Busch and Sievers 1990). Moreover, mutant analyses also confirmed a connection between starch content of amyloplasts and gravitropic responses. Mutants of *PHOSPHOGLUCOMUTASE* (*PGM*) in *A. thaliana* and *Nicotiana sylvestris* that form starchless plastids in columella cells are impaired in gravitropism, highlighted by a delayed and weak gravitropic response (Kiss and Sack 1989; Kiss et al. 1989). It was assumed that a reduction in the mass of the amyloplasts led to a delayed sedimentation and thereby to a delayed graviperception (Kiss and Sack 1989). This was confirmed by experiments with the *Arabidopsis thaliana* mutant *starch excess 1* (*sex1*) which has amyloplasts with higher starch content in the hypocotyl and requires a shorter presentation time before bending can be observed (Vitha et al. 2007).

Auxin redistribution in response to plant reorientation is also severely disrupted in the *pgm 1* mutant as shown by the auxin sensor DII-Venus (Band et al. 2012). The process of statolith sedimentation therefore appears to be fundamental for the perception of gravity.

The starch-statolith hypothesis includes all perception mechanisms that are driven by the sedimentation of statoliths. It is based on mechanisms in which the sedimentation of statoliths ultimately leads to a contact with or a movement along the ER or plasma membrane and thereby to perception of the biophysical stimulus. This can be achieved by the opening of mechanosensitive channels or via protein-protein interaction, which can both trigger a biochemical signaling cascade. Statolith sedimentation exerts pressure on the ER or plasma membrane. A proportion of amyloplasts is usually already in close proximity to the distal ER and moves along the ER upon gravistimulation. Both movement and pressure can cause membrane deformation after sedimentation (Behrens et al. 1985; Leitz et al. 2009), which yields the potential to open membrane-localized mechanosensitive ion channels (Hamilton et al. 2015). The resulting ion current can trigger further signaling cascades (Toyota and Gilroy 2013). Calcium ions are suspected to be the ions that are transported, as the ER is a prominent calcium store (Meldolesi and Pozzan 1998). Other ions that can pass mechanosensitive channels are potassium and chloride (Toyota and Gilroy 2013). The identity of the suspected mechanosensitive channels has not been revealed yet (reviewed by Baldwin et al. 2013). A binding of receptor (in the ER membrane) and ligand (in the statolith membrane) could also trigger gravitropic signaling cascades. Experiments with *Chara* rhizoids offer results that could in part also be true for higher plants (see Chap. 4). A transfer of results obtained with *Chara* to higher plants, however, must consider that statocytes have a different architecture. *Chara* rhizoids do not have a peripheral positioned ER. Contact between statolith and membrane in *Chara* takes place with the plasma membrane, in higher plants with the ER. If protein-protein interactions play a role in gravitropism in higher plants, an evolutionary relocalization of proteins or new receptor ligand complexes would have to be assumed.

It is worth noting that there is a significant difference in timing between the duration of statolith sedimentation and first measurable responses. Complete sedimentation of statolith was reported to take about 5–10 min (Leitz et al. 2009;

Baldwin et al. 2013). However, with intermittent stimulation of cress roots on a clinostat, the perception time was calculated to be in a range of 0.5 s, which strongly indicates that a minor movement of statoliths already settled on ER-cisternae is sufficient for graviperception to occur (Heinowicz et al. 1998). Measurements of membrane potentials following a 5–15 s reorientation show membrane depolarizations already 8 s after reorientation (Behrens et al. 1985). An increase in cytoplasmic inositol-1,4,5-triphosphate (InsP$_3$) was observed in *Zea mays* 10 s after reorientation (Perera et al. 1999). Measurements of cytosolic calcium in parabolic flight experiments also suggest changes within seconds of the microgravity stimulus (Neef et al. 2016).

6.2.3 Tensegrity Model and the Role of the Actin in Gravity Sensing

Tensegrity models describe the connection between graviperception and the cytoskeleton of the cell. Tensegrity is a composite of the words tension and integrity and describes biologically the connection between mechanical force and mechanisms or functions of the cell (Ingber et al. 2014). Statoliths are surrounded by an only very delicate actin microfilament meshwork (Collings et al. 2001). First tensegrity models described that the actin cytoskeleton surrounding the statoliths is connected to statolith and ER cisternae or the plasma membrane. Sedimentation of a statolith would put tension on the cytoskeleton and open mechanosensitive channels in the plasma membrane (Sievers et al. 1989). One prediction from this model was that a disruption of the actin cytoskeleton would prevent graviperception. But in fact, experiments in which the actin filaments were disrupted by Latrunculin B did not lead to an agravitropic phenotype and application of Latrunculin B and Cytochalasin B, another actin depolymerizing drug, even increased statolith sedimentation velocity and gravitropic responses in cress roots (Sievers et al. 1989) and in *Zea mays* roots (Blancaflor 2002; Hou et al. 2003, 2004). Furthermore, the promoted graviresponse that was found in roots, which were gravistimulated for only a short time and then rotated on a clinostat, was attributed to the inability of fragmented actin microfilaments to terminate the curvature response (Hou et al. 2004). Thus, the actin cytoskeleton likely does not play a crucial role in the gravity-sensing process and it is tempting to speculate that actin is not essential for gravity sensing *per se* but is required for controlling and fine-tuning an appropriate and functional resting position as well as unimpeded sedimentation of statoliths. In fact, actomyosin forces may even increase the energetic noise level of the sensing mechanism by increasing the random movements of statoliths, making statocytes less susceptible for unfavorable fast responses to quickly changing or only transient gravistimulation, e.g. when the wind repeatedly bends crop stalks (Braun and Limbach 2006).

6.3   Signal Transduction and Transmission

It is obvious that a mode of communication exists between the root tip cells that sense gravity and the cells in the elongation zone that respond to the gravity signal by expanding differentially on the opposite root flanks which leads to the downward curvature. After conversion of the physical stimulus into a biochemical signal (signal transduction) in the columella cells, the signal of all statocytes must be integrated, enhanced, transmitted and received in the responding tissue.

6.3.1   Secondary Messengers

Signal transduction and transmission employs a plethora of secondary messengers including calcium ions, InsP$_3$ and protons (Fasano 2001; Perera 2006).

Calcium is an essential secondary messenger in many signaling processes in plants and has been considered as an important messenger in gravitropism. It was noted early on that radioactively labeled $^{45}Ca^{2+}$ is transported basally in *Zea mays* roots after gravistimulation (Lee and Evans 1985). One-sided application of calcium to the root leads to bending towards the source of calcium (Lee et al. 1983a). Gravitropic bending is severely impaired by application of calcium chelators or by inhibition of calmodulin or calcium-channels (Lee et al. 1983b; Vanneste and Friml 2013). These observations suggest that calcium plays a role in the apoplast of the root (Toyota and Gilroy 2013). Calmodulins, prominent calcium binding proteins, are highly expressed in the root tip (Stinemetz et al. 1987), which might suggest a role of calcium also in gravity perception.

More recent experiments with the calcium sensor aequorin support a role for calcium during early gravitropic responses (Toyota et al. 2007). Aequorin fluorescence is dependent on the availability of intracellular calcium (Shimomura et al. 1962). After reorientation experiments, two calcium waves were measured in statocytes of the hypocotyl, the first after 4 s, the second after 1 min. The first calcium wave was considered an effect of the mechanical reorientation, the second as an effect of the gravitational response. It remains unclear whether calcium is part of the signal perception or part of the later signaling events. The second wave suggests an involvement in signal transduction and not signal perception. This is supported by physiological changes, like membrane depolarization, that appear earlier than 1 min (Behrens et al. 1985). More recently introduced calcium sensors, like Cameleon YC3.6, have been used to detect a calcium wave arising on the lower side of the root and moving towards the elongation zone like auxin (Monshausen et al. 2011). Calcium could play different roles in early as well as progressed signal transduction. Modern real-time measurements, e.g. in parabolic plane flights, promise more detailed results in the coming years (Neef et al. 2016).

Calcium intimately interplays with auxin. Auxin induces calcium levels and signaling (Toyota and Gilroy 2013). Increases in calcium regulate auxin transport

(Robert and Offringa 2008). Calcium elevations change cell wall pH which regulates elongation via acid growth (Monshausen et al. 2011). The proton concentration (pH) of a cellular compartment has a strong effect on all containing proteins and metabolites. A change in pH can have a modulating influence on various cellular activities. Using a pH sensor, a movement of protons from the cytoplasm to the apoplast was observed in the columella in *Arabidopsis thaliana* upon gravistimulation (Fasano 2001). Alkalization of the cytoplasm occurs within the first 2 min with a change from pH 7.2 to pH 7.6, while the pH in the apoplast changes from pH 5.5 to 4.5. Later pH changes in the cell wall in the elongation zone were attributed to tropic growth. Similar observations of pH changes were made in *Zea mays* pulvini in the statocytes-containing endodermis (Johannes et al. 2001). With the help of caged protons that were released upon UV irradiation, the beginning of tropic growth could be delayed by manual manipulation of the pH value (Fasano 2001).

Mutants defective in columella cell alkalization fail to relocate auxin transporters of the PIN family, thereby affecting auxin redistribution (Boonsirichai et al. 2003; Harrison and Masson 2008).

Phosphatidylinositol-4,5-bisphosphate (PIP$_2$) is cleaved via hydrolysis by phospholipase C into diacylglycerol and InsP$_3$. Both hydrolysis products play a role in downstream signaling cascades. Mutants of *Phosphatidylinositol-monophosphate-5-kinase* (*PIP5K*) are delayed in their gravitropic response and impaired in polar auxin transport (Mei et al. 2012). One of the products of PIP5K, InsP$_3$ modulates intracellular calcium signals in animal cells. InsP$_3$ opens ER-localized calcium channels which trigger downstream signaling cascades (Berridge 2009). InsP$_3$ could have a similar influence on calcium signaling in plants during gravitropism. A role for InsP$_3$ in gravitropism is supported by several experiments. A fivefold increase in InsP$_3$ concentration in the lower pulvinus, the area of bending growth in *Zea mays*, was measured 10 s after reorientation (Perera et al. 1999). The increase in InsP$_3$ correlates with membrane depolarization after 8 s (Behrens et al. 1985) and suggests that InsP$_3$ plays a role during early gravitropism signaling or even graviperception. The InsP$_3$ concentration is still increased after 2 hours, suggesting an involvement also in later signaling processes. Following the gravitropic response (8–10 h), the InsP$_3$ concentration reverts to its initial level. In plants overexpressing human INOSITOL-POLYPHOSPHATE-5-PHOSPHATASE, catalyzing constitutive hydrolysis of InsP$_3$, gravitropic responses were present but reduced (Perera 2006). Inhibition of InsP$_3$ synthesis by blocking PHOSPHOLIPASE C in *Arabidopsis thaliana* leads to reduced gravitropic responses in roots and shoot (Andreeva et al. 2010).

It is noteworthy that gravitropic responses in InsP$_3$-deficient plants were never abolished but always reduced. InsP$_3$ can carefully be considered modulating but not essential for gravitropism. InsP$_3$ may have to be considered in the context of other secondary messengers. A direct connection between InsP$_3$ and calcium and InsP$_3$-activated calcium channels is a point of future research.

While calcium, InsP$_3$, and pH play a role as secondary messengers in gravitropism signaling, the resulting hormone distribution leads to the final curvature responses.

6.3.2 Asymmetric Hormone Distribution Leads to Directed Growth

Cholodny and Went independently discovered a connection between phytohormones and tropisms in higher plants. Cholodny observed that a gravitational stimulus disturbs the even distribution of phytohormones in the root of higher plants. Went also observed that the phytohormone auxin specifically enriches on the bending side. Later experiments on the influence of light on hormone distribution led to similar results (Went and Thimann 1937). Based on these observations the Cholodny-Went hypothesis was postulated that claims that the bending growth typical for tropisms is a result of polar auxin distribution (Cholodny 1929).

Later, the Cholodny-Went hypothesis was refined by the observation that besides auxin, the phytohormone jasmonic acid (JA) also shows a polar distribution. Inhibition of the JA gradient led to a delay of gravitropic responses. JA-deficient *Oryza sativa* mutants show delayed but observable gravitropic responses (Gutjahr et al. 2005). This suggests a modulating function of JA in gravitropism.

A central element of the gravitropic growth response is, therefore, the generation of hormone gradients within the growing organ. For auxin this is achieved via carrier-mediated asymmetric transport.

6.4 Gravitropic Growth

6.4.1 Polar Hormone Distribution

While the importance of auxin for tropic growth was known since the Cholodny-Went hypothesis, the mechanism of auxin transport was only identified later. The most important member of the group of auxins, free Indole-3-acetic acid can be protonated (IAAH) or deprotonated (IAA$^-$) depending on the pH. Only IAAH can freely diffuse through the plasma membrane. At pH 7 IAA$^-$ prevails and requires active transport from cell to cell (Friml and Palme 2002).

Experiments with auxin transport inhibitors led to different results in Boston Ivy (*Parthenocissus tricuspidata*). Triiodobenzoic acid (TIBA) inhibits auxin efflux from the cytoplasm. Auxin accumulation in the cytoplasm was still observed besides TIBA application. 2,4-Dichlorphenoxyacetic acid inhibits auxin influx. Based on these observations, the chemiosmotic theory was postulated. The interplay of passive auxin transport with the activity of auxin efflux and influx carriers allows polar auxin transport in plants (Rubery and Sheldrake 1974).

Modern molecular biology methods allowed for the identification of the efflux and influx carriers. Mutant analyses identified *AUXIN RESISTANT 1 (AUX1)* as the gene coding for the influx carrier (Bennett et al. 1996). Similar mutant experiments identified *PIN-FORMED 1 (PIN1)* as a gene coding for an efflux carrier, with polar localization in the plasma membrane and thereby allowing for polar auxin transport

(Gälweiler et al. 1998). Further PIN proteins and their subcellular localization were identified. In general, expression and localization of PIN proteins is cell-type dependent (Michniewicz et al. 2007). Further efflux carriers encompass proteins of the *ABC-B/MULTI-DRUG RESISTANCE/P-GLYCOPROTEIN (ABCB/MDR/PGP)* family. They are localized symmetrically in the plasma membrane and are involved in auxin homeostasis (Cho and Cho 2013).

The polar auxin transport must work more efficiently on one side of the gravistimulated organ after reorientation of the plant, so that according to Cholodny and Went an asymmetric auxin distribution can occur. As a result, stronger growth on one side will lead to bending of the plant. It was known that the cell polarity of PIN localization depends on cell type and developmental stage of the tissue and therefore allows directed auxin flow. PIN1 and PIN7 have different polarized localizations in pro-embryo and adult plant (Friml 2003). For PIN3 it was shown that gravistimulation has an influence on the localization of the efflux carrier. PIN3 relocalization in columella cells always appears at the new physiological bottom (Friml et al. 2002). An accumulation of PIN3 at the new physiological bottom would favor auxin transport along the new lower side of the root and lead to root bending.

PIN proteins are localized via vesicular transport. An important component of these vesicular transport systems are ENDOSOMAL SORTING COMPLEXES REQUIRED FOR TRANSPORT (ESCRT) that are formed by various VESICU-LAR SORTING PROTEINS (VSPs). Mutations in the regulatory system of ESCRT lead to different localizations of PIN1, PIN2 and AUX1. A double mutant of CHARGED MULTIVESICULAR BODY PROTEIN/CHROMATIN MODIFY-ING PROTEIN1A (CHMP1A) and CHMP1B leads to accumulation of AUX1, PIN1 and PIN2 in late endosomes and only marginal localization at the plasma membrane (Spitzer et al. 2009).

6.5   Microgravity Research and Modifying Gravitational Acceleration Changed Our Perspective on Gravitropism

Centrifugation has been widely used to alter the amount and the direction of mass acceleration that acts on the sedimentable masses in gravity sensing systems. In a pioneering experiment, Sir Thomas Andrew Knight used a water-driven horizontal wheel to show that plant roots neither grow in the direction of gravity nor in the direction of the centrifugal acceleration but grow in the direction of the resulting acceleration angle, thus, providing evidence that plants sense the direction of accelerations not gravity *per se* (Knight 1806). Centrifugation intensifies weak gravity responses (improved graviorientation in starchless *Arabidopsis* mutants), renders intracellular processes more clearly visible and was used to characterize gravisensitive membranes by moving statoliths (cf. Chap. 4). Very high acceleration forces were even used to reversibly abolish graviresponsiveness by disturbing the polar cytoplasmic organization of statocytes (Wendt and Sievers 1986).

Since Julius Sachs made his first attempts to neutralize the unilateral effect of gravity by rotating plants around a horizontal axis (Sachs 1882), numerous types of clinostats have been developed in order to simulate the effects of weightlessness on Earth (cf. Chap. 2). Classical clinostats rotating with 1–10 rpm, fast-rotating clinostats rotating with 50–120 rpm as well as three-dimensional clinostats and random positioning machines rotating around one axis or two axes have shed light on the gravisensitivity, perception and presentation time of gravitropism in roots and shoots.

A new quality of research tools for studies on gravitropism became available with the advent of drop towers, parabolic flights of aircrafts and rockets, NASAs Space Shuttles, satellites and Space Stations in the low Earth orbit (cf. Chap. 2). With the completion of the International Space Station ISS with the commissioning of the Columbus module in 2008, the biggest microgravity lab that ever existed provides an almost stimulus-free environment of real microgravity for long-term experimentation. The microgravity quality in a range of 10^{-3}–10^{-6} g is beyond the susceptibility of most biological sensory systems and is, therefore, regarded as functional weightlessness allowing biologists to address especially molecular and cellular mechanisms involved in plant gravity sensing and graviorientation. Only after cress seeds had been germinated on the Space Shuttle (Volkmann et al. 1986) and on Russian Bion satellites (Laurinavicius et al. 1996), it became evident that the development of polarly organized statocytes and gravisensing mechanisms are neither induced nor affected by the absence of gravity. And although a reduced starch content was reported, microgravity-grown roots responded more strongly to only small acceleration doses in microgravity aboard Space Shuttles (Volkmann and Tewinkel 1996; Perbal et al. 2004). The presentation time, the time a stimulus needs to be applied continuously to a sensing system in order to trigger a response, was found to be in the range of 20–30 s when stimulated with normal 1 g acceleration, whereas the presentation time of cress and lentil roots grown on a 1 g centrifuges was in a range of 50–60 s (Perbal and Driss-Ecole 1994; Volkmann and Tewinkel 1996). By rotating roots on a clinostat and stopping it several times, a perception time of 1 s was determined for cress roots. The perception time defines the minimum time a stimulus is registered by sensing systems but must be given repeatedly in order to trigger a response (Heinowicz et al. 1998). In this short period, statoliths in gravistimulated roots are displaced only a fraction of a µm, which is good evidence that—taken into account that actin is not required for graviperception as was demonstrated by the uninhibited graviresponse of roots with disrupted actin microfilaments (Hou et al. 2003)—graviperception must occur very close to statoliths already sedimented on or in close contact with a gravisensitive endoplasmic reticulum membrane.

Clinorotating seedling has been used very often to successfully prevent static gravitropic stimulation of roots by randomizing the gravity vector and, consequently, roots continued to grow straight. However, clinorotation failed in most cases to eliminate also dynamic stimulation. There are studies reporting an overloading of the sensory system due to vibrations and shifting statoliths continuously, which could even result in the complete disintegration of the polar structural organization of the statocytes after hours of clinorotation (Hensel and Sievers 1980; Hoson et al. 1997). Therefore, research focusing on specific molecular and cellular components of gravity sensing mechanisms and on the role of second messengers like cytoplasmic free calcium, InsP3, and pH in gravity signaling pathways are best performed under real-microgravity conditions in space. Recently, experiments on the ISS have provided evidence that the establishment of the auxin-gradient system, the prerequisite for the curvature response, is gravity independent. The cytokinin distribution, however, was different in space-grown and control roots suggesting that cytokinin-associated process involved in gravitropism might be affected (Ferl and Paul 2016).

Further research on the ISS and other microgravity platforms is currently planned to increase our knowledge on the hormone-associated components of gravitropic responses. In particular future research on mutants in the stimulus-free microgravity environment promises to further contribute to the unravelling of the fascinating processes of gravity sensing and gravitropic orientation in higher plants.

References

Andreeva Z, Barton D, Armour WJ et al (2010) Inhibition of phospholipase C disrupts cytoskeletal organization and gravitropic growth in *Arabidopsis* roots. Planta 232:1263–1279. https://doi.org/10.1007/s00425-010-1256-0

Audus L (1979) Plant geosensors. J Exp Bot 8:235–249

Baldwin KL, Strohm AK, Masson PH (2013) Gravity sensing and signal transduction in vascular plant primary roots. Am J Bot 100:126–142. https://doi.org/10.3732/ajb.1200318

Baluška F, Kreibaum A, Vitha S et al (1997) Central root cap cells are depleted of endoplasmic microtubules and actin microfilament bundles: implications for their role as gravity-sensing statocytes. Protoplasma 196:212–223. https://doi.org/10.1007/BF01279569

Band LR, Wells DM, Larrieu A et al (2012) Root gravitropism is regulated by a transient lateral auxin gradient controlled by a tipping-point mechanism. Proc Natl Acad Sci 109:4668–4673. https://doi.org/10.1073/pnas.1201498109

Behrens HM, Gradmann D, Sievers A (1985) Membrane-potential responses following gravistimulation in roots of *Lepidium sativum* L. Planta 163:463–472. https://doi.org/10.1007/BF00392703

Bennett MJ, Marchant A, Green HG et al (1996) Arabidopsis AUX1 gene: a permease-like regulator of root gravitropism. Science 273:948–950. https://doi.org/10.1126/science.273.5277.948

Berridge MJ (2009) Inositol trisphosphate and calcium signalling mechanisms. Biochim Biophys Acta, Mol Cell Res 1793:933–940

Blancaflor EB (2002) The cytoskeleton and gravitropism in higher plants. J Plant Growth Regul 21:120–136. https://doi.org/10.1007/s003440010041

Blancaflor EB, Fasano JM, Gilroy S (1998) Mapping the functional roles of cap cells in the response of *Arabidopsis* primary roots to gravity. Plant Physiol 116:213–222. https://doi.org/10.1104/pp.116.1.213

Boonsirichai K, Sedbrook JC, Chen R et al (2003) ALTERED RESPONSE TO GRAVITY is a peripheral membrane protein that modulates gravity-induced cytoplasmic alkalinization and lateral auxin transport in plant statocytes. Plant Cell 15:2612–2625

Braun M, Limbach C (2006) Rhizoids and protonemata of characean algae: model cells for research on polarized growth and plant gravity sensing. Protoplasma 229:133–142. https://doi.org/10.1007/s00709-006-0208-9

Busch MB, Sievers A (1990) Hormone treatment of roots causes not only a reversible loss of starch but also of structural polarity in statocytes. Planta 181:358–364. https://doi.org/10.1007/BF00195888

Cho M, Cho HT (2013) The function of ABCB transporters in auxin transport. Plant Signal Behav 8:2–4. https://doi.org/10.4161/psb.22990

Cholodny N (1929) Einige Bemerkungen zum Problem der Tropismen. Planta 7:461–481. https://doi.org/10.1007/BF01912159

Collings DA, Zsuppan G, Allen NS, Blancaflor EB (2001) Demonstration of prominent actin filaments in the root columella. Planta 212:392–403. https://doi.org/10.1007/s004250000406

Darwin CR (1880) The power of movement in plants. John Murray, London

Digby J, Firn RD (1995) The gravitropic set-point angle (GSA): the identification of an important developmentally controlled variable governing plant architecture. Plant Cell Environ 18:1434–1440. https://doi.org/10.1111/j.1365-3040.1995.tb00205.x

Fasano JM (2001) Changes in root cap pH are required for the gravity response of the *Arabidopsis* root. Plant Cell 13:907–922. https://doi.org/10.1105/tpc.13.4.907

Ferl RJ, Paul A-L (2016) The effect of spaceflight on the gravity-sensing auxin gradient of roots: GFP reporter gene microscopy on orbit. NPJ Microgravity 2:15023. https://doi.org/10.1038/npjmgrav.2015.23

Frank AB (1868) Über die durch die Schwerkraft verursachten Bewegungen von Pflanzentheilen. Beiträge zur Pflanzenphysiologie 8:1–99

Friml J (2003) Auxin transport – shaping the plant. Curr Opin Plant Biol 6:7–12. https://doi.org/10.1016/S1369-5266(02)00003-1

Friml J, Palme K (2002) Polar auxin transport – old questions and new concepts? Plant Mol Biol 49:273–284. https://doi.org/10.1007/978-94-010-0377-3_2

Friml J, Wiśniewska J, Benková E et al (2002) Lateral relocation of auxin efflux regulator PIN3 mediates tropism in *Arabidopsis*. Nature 415:806–809. https://doi.org/10.1038/415806a

Fukaki H, Wysocka-Diller J, Kato T et al (1998) Genetic evidence that the endodermis is essential for shoot gravitropism in *Arabidopsis thaliana*. Plant J 14:425–430. https://doi.org/10.1046/j.1365-313X.1998.00137.x

Gälweiler L, Guan C, Müller A et al (1998) Regulation of polar auxin transport by AtPIN1 in *Arabidopsis* vascular tissue. Science 282:2226–2230. https://doi.org/10.1126/science.282.5397.2226

Gutjahr C, Riemann M, Müller A et al (2005) Cholodny-went revisited: a role for jasmonate in gravitropism of rice coleoptiles. Planta 222:575–585. https://doi.org/10.1007/s00425-005-0001-6

Guyomarc'h S, Leran S, Auzon-Cape M et al (2012) Early development and gravitropic response of lateral roots in *Arabidopsis thaliana*. Philos Trans R Soc B Biol Sci 367:1509–1516. https://doi.org/10.1098/rstb.2011.0231

Haberlandt G (1900) Über die Perzeption des geotropischen Reizes. Ber Dtsch Bot Ges 18:261–272

Hamilton ES, Schlegel AM, Haswell ES (2015) United in diversity: mechanosensitive ion channels in plants. Annu Rev Plant Biol 66:113–137. https://doi.org/10.1146/annurev-arplant-043014-114700

Harrison BR, Masson PH (2008) ARL2, ARG1 and PIN3 define a gravity signal transduction pathway in root statocytes. Plant J 53:380–392. https://doi.org/10.1111/j.1365-313X.2007.03351.x

Heinowicz Z, Sondag C, Alt W, Sievers A (1998) Temporal course of graviperception in intermittently stimulated cress roots. Plant Cell Environ 21:1293–1300. https://doi.org/10.1046/j.1365-3040.1998.00375.x

Hensel W, Sievers A (1980) Effects of prolonged omnilateral gravistimulation on the ultrastructure of statocytes and on the graviresponse of roots. Planta 150:338–346. https://doi.org/10.1007/BF00384664

Hoson T, Kamisaka S, Masuda Y et al (1997) Evaluation of the three-dimensional clinostat as a simulator of weightlessness. Planta 203:S187–S197. https://doi.org/10.1007/PL00008108

Hou G, Mohamalawari DR, Blancaflor EB (2003) Enhanced gravitropism of roots with a disrupted cap actin cytoskeleton. Plant Physiol 131:1360–1373. https://doi.org/10.1104/pp.014423

Hou G, Kramer VL, Wang YS et al (2004) The promotion of gravitropism in Arabidopsis roots upon actin disruption is coupled with the extended alkalinization of the columella cytoplasm and a persistent lateral auxin gradient. Plant J 39:113–125. https://doi.org/10.1111/j.1365-313X.2004.02114.x

Ingber DE, Wang N, Stamenović D (2014) Tensegrity, cellular biophysics, and the mechanics of living systems. Reports Prog Phys 77. https://doi.org/10.1088/0034-4885/77/4/046603

Johannes E, Collings DA, Rink JC, Allen NS (2001) Cytoplasmic pH dynamics in maize pulvinal cells induced by gravity vector changes. Plant Physiol 127:119–130. https://doi.org/10.1104/pp.127.1.119

Juniper BE, Groves S, Landau-Schachar B, Audus LJ (1966) Root cap and the perception of gravity [35]. Nature 209:93–94

Kiss JZ, Sack FD (1989) Reduced gravitropic sensitivity in roots of a starch-deficient mutant of Nicotiana sylvestris. Planta 180:123–130. https://doi.org/10.1007/BF02411418

Kiss JZ, Hertel R, Sack FD (1989) Amyloplasts are necessary for full gravitropic sensitivity in roots of Arabidopsis thaliana. Planta 177:198–206. https://doi.org/10.1007/BF00392808

Knight TA (1806) V. On the direction of the radicle and germen during the vegetation of seeds. By Thomas Andrew knight, Esq. F. R. S. In a letter to the right Hon. Sir Joseph banks, K. B. P. R. S. Philos Trans R Soc London 96:99–108. https://doi.org/10.1098/rstl.1806.0006

Konings H (1968) The significance of the root cap for geotropism. Acta Bot Neerl 17:203–211. https://doi.org/10.1111/j.1438-8677.1968.tb00074.x

Kuznetsov OA, Hasenstein KH (1996) Intracellular magnetophoresis of amyloplasts and induction of root curvature. Planta 198:87–94. https://doi.org/10.1007/BF00197590

Kuznetsov OA, Hasenstein KH (1997) Magnetophoretic induction of curvature in coleoptiles and hypocotyls. J Exp Bot 48:1951–1957. https://doi.org/10.1093/jexbot/48.316.1951

Kuznetsov OA, Schwuchow J, Sack FD, Hasenstein KH (1999) Curvature induced by amyloplast magnetophoresis in protonemata of the moss Ceratodon purpureus. Plant Physiol 119 (2):645–650

Laurinavicius R, Stockus A, Buchen B, Sievers A (1996) Structure of cress root statocytes in microgravity (Bion-10 mission). Adv Space Res 17:91–94

Lee JS, Evans ML (1985) Polar transport of auxin across gravistimulated roots of maize and its enhancement by calcium. Plant Physiol 77:824–827

Lee JS, Mulkey TJ, Evans ML (1983a) Gravity-induced polar transport of calcium across root tips of maize. Plant Physiol 73:874–876. https://doi.org/10.1104/pp.73.4.874

Lee JS, Mulkey TJ, Evans ML (1983b) Reversible loss of gravitropic sensitivity in maize roots after tip application of calcium chelators. Science 220:1375–1376

Leitz G, Kang B-H, Schoenwaelder MEA, Staehelin LA (2009) Statolith sedimentation kinetics and force transduction to the cortical endoplasmic reticulum in gravity-sensing Arabidopsis columella cells. Plant Cell Online 21:843–860. https://doi.org/10.1105/tpc.108.065052

Mei Y, Jia WJ, Chu YJ, Xue HW (2012) Arabidopsis phosphatidylinositol monophosphate 5-kinase 2 is involved in root gravitropism through regulation of polar auxin transport by affecting the cycling of PIN proteins. Cell Res 22:581–597. https://doi.org/10.1038/cr.2011.150

Meldolesi J, Pozzan T (1998) The endoplasmic reticulum Ca^{2+} store: a view from the lumen. Trends Biochem Sci 23:10–14

Michniewicz M, Zago MK, Abas L et al (2007) Antagonistic regulation of PIN phosphorylation by PP2A and PINOID directs auxin flux. Cell 130:1044–1056. https://doi.org/10.1016/j.cell.2007.07.033

Monshausen GB, Miller ND, Murphy AS, Gilroy S (2011) Dynamics of auxin-dependent Ca^{2+}and pH signaling in root growth revealed by integrating high-resolution imaging with automated computer vision-based analysis. Plant J 65:309–318. https://doi.org/10.1111/j.1365-313X.2010.04423.x

Mullen JL, Hangarter RP (2003) Genetic analysis of the gravitropic set-point angle in lateral roots of Arabidopsis. Adv Space Res 31:2229–2236

Neef M, Denn T, Ecke M, Hampp R (2016) Intracellular calcium decreases upon hyper gravity-treatment of *Arabidopsis thaliana* cell cultures. Microgravity Sci Technol 28:331–336. https://doi.org/10.1007/s12217-015-9457-6

Nemec B (1900) Ueber die Art der Wahrnehmung des Schwerkraftreizes bei den Pflanzen. Ber Dtsch Bot Ges 18:241–245

Palmieri M, Kiss JZ (2007) The role of plastids in gravitropism. In: The structure and function of plastids. Springer, Dordrecht, pp 507–525

Perbal G, Driss-Ecole D (1994) Sensitivity to gravistimulus of lentil seedling roots grown in space during the IML 1 mission of spacelab. Physiol Plant 90:313–318. https://doi.org/10.1111/j.1399-3054.1994.tb00393.x

Perbal G, Lefranc A, Jeune B, Driss-Ecole D (2004) Mechanotransduction in root gravity sensing cells. Physiol Plant 120:303–311. https://doi.org/10.1111/j.0031-9317.2004.0233.x

Perera IY (2006) A universal role for inositol 1,4,5-trisphosphate-mediated signaling in plant gravitropism. Plant Physiol 140:746–760. https://doi.org/10.1104/pp.105.075119

Perera IY, Heilmann I, Boss WF (1999) Transient and sustained increases in inositol 1,4,5-trisphosphate precede the differential growth response in gravistimulated maize pulvini. Proc Natl Acad Sci U S A 96:5838–5843. https://doi.org/10.1073/pnas.96.10.5838

Pfeffer W (1904) Pflanzenphysiologie: ein Handbuch der Lehre vom Stoffwechsels und Kraftwechsels in der Pflanze. W. Engelmann, Leipzig

Robert HS, Offringa R (2008) Regulation of auxin transport polarity by AGC kinases. Curr Opin Plant Biol 11:495–502

Rubery PH, Sheldrake AR (1974) Carrier-mediated auxin transport. Planta 118:101–121. https://doi.org/10.1007/BF00388387

Sachs J (1882) Über Ausschließung der geotropischen und heliotropischen Krümmungen während des Wachsens. Wilhelm Engelmann, Leipzig

Schüler O, Hemmersbach R, Böhmer M (2015) A bird's-eye view of molecular changes in plant gravitropism using omics techniques. Front Plant Sci 6. https://doi.org/10.3389/fpls.2015.01176

Shimomura O, Johnson FH, Saiga Y (1962) Extraction, purification and properties of aequorin, a bioluminescent protein from the luminous gydromedusan, *Aequorea*. J Cell Comp Physiol 59:223–239. https://doi.org/10.1002/jcp.1030590302

Sievers A, Volkmann D (1971) Verursacht differentieller Druck der Amyloplasten auf ein komplexes Endomembransystem die Geoperzeption in Wurzeln? Planta 102:160–172. https://doi.org/10.1007/BF00384870

Sievers A, Volkmann D (1977) Ultrastructure of gravity-perceiving cells in plant roots. Proc R Soc Lond B 199:525–536. https://doi.org/10.1098/rspb.1977.0160

Sievers A, Kruse S, Kuo-Huang LL, Wendt M (1989) Statoliths and microfilaments in plant cells. Planta 179:275–278. https://doi.org/10.1007/BF00393699

Spitzer C, Reyes FC, Buono R et al (2009) The ESCRT-related CHMP1A and B proteins mediate multivesicular body sorting of auxin carriers in *Arabidopsis* and are required for plant development. Plant Cell Online 21:749–766. https://doi.org/10.1105/tpc.108.064865

Stinemetz CL, Kuzmanoff KM, Evans ML, Jarrett HW (1987) Correlation between calmodulin activity and gravitropic sensitivity in primary roots of maize. Plant Physiol 84:1337–1342. https://doi.org/10.1104/pp.84.4.1337

Toyota M, Gilroy S (2013) Gravitropism and mechanical signaling in plants. Am J Bot 100:111–125. https://doi.org/10.3732/ajb.1200408

Toyota M, Furuichi T, Tatsumi H, Sokabe M (2007) Hypergravity stimulation induces changes in
    intracellular calcium concentration in *Arabidopsis* seedlings. Adv Space Res 39:1190–1197.
    https://doi.org/10.1016/j.asr.2006.12.012
Tsugeki R, Fedoroff NV (1999) Genetic ablation of root cap cells in *Arabidopsis*. Proc Natl Acad
    Sci U S A 96:12941–12946. https://doi.org/10.1073/pnas.96.22.12941
Vanneste S, Friml J (2013) Calcium: the missing link in auxin action. Plants 2:650–675
Vitha S, Yang M, Sack FD, Kiss JZ (2007) Gravitropism in the starch excess mutant of *Arabidopsis
    thaliana*. Am J Bot 94:590–598. https://doi.org/10.3732/ajb.94.4.590
Volkmann D, Tewinkel M (1996) Gravisensitivity of cress roots: investigations of threshold values
    under specific conditions of sensor physiology in microgravity. Plant Cell Environ
    19:1195–1202. https://doi.org/10.1111/j.1365-3040.1996.tb00435.x
Volkmann D, Behrens HM, Sievers A (1986) Development and gravity sensing of cress roots under
    microgravity. Naturwissenschaften 73:438–441. https://doi.org/10.1007/BF00367291
Wayne R, Staves MP (1996) A down to earth model of gravisensing or Newton's law of gravitation
    from the apple's perspective. Physiol Plant 98:917–921. https://doi.org/10.1111/j.1399-3054.
    1996.tb06703.x
Wendt M, Sievers A (1986) Restitution of polarity in statocytes from centrifuged roots. Plant Cell
    Environ 9:17–23. https://doi.org/10.1111/1365-3040.ep11612684
Went FW, Thimann KV (1937) Phytohormones. The Macmillan Company, New York
Yoder TL (2001) Amyloplast sedimentation dynamics in maize columella cells support a new
    model for the gravity-sensing apparatus of roots. Plant Physiol 125:1045–1060. https://doi.org/
    10.1104/pp.125.2.1045
Zheng HQ, Staehelin LA (2001) Nodal endoplasmic reticulum, a specialized form of endoplasmic
    reticulum found in gravity-sensing root tip columella cells. Plant Physiol 125:252–265. https://
    doi.org/10.1104/pp.125.1.252

Chapter 7
Gravitropism in Higher Plants: Molecular Aspects

Klaus Palme, William Teale, and Franck Ditengou

Abstract The pervasive influence of gravity on life on Earth presents barriers to our identifying and understanding of the signaling pathways which have evolved in response to it. Plants are at the same time positively and negatively gravitropic, using the Earth's gravity to define their stature both above and below ground. Here we review some of the signaling pathways which use the plant hormone auxin to carry information on orientation from regions of perception to regions of growth response. The regulation of these pathways is at once diverse and as yet poorly understood but involves the control of members of a family of polarly localized cellular auxin efflux carriers, the PINs, by factors such as phosphorylation. Auxin transport is also influenced by the availability of calcium ions; this interaction is likely to emerge as a key node in a plant's responses to gravity. It is hoped that understanding the mechanism of these responses will not only allow more efficient cultivation of plants in space, but open paths to greater control over plant stature which will enable us, in the future, better to respond to the challenges of feeding those of us still living on Earth.

Keywords Auxin · Higher plant gravitropism · Kinase signaling · Microgravity · Plant hormone

7.1 Introduction

On spaceship Earth, all evolution is governed by a 1-g environment. Different organisms have evolved diverse strategies to monitor the gravitational field and use it as a positional cue to orientate their growth. In this, plants are no exception: they are sessile, but use gravity as a signal which integrates with other inputs (such as light intensity and direction, humidity, touch and temperature), coordinating growth to optimize access to light, water and nutrients. It is not surprising that plants are a long-standing and important target for research into the mechanisms underlying gravity perception and the gravity response. While cellular aspects of gravitropism are discussed in Chap. 6, here we will discuss molecular aspects mostly unraveled in *Arabidopsis thaliana*. The use of this fully sequenced and well characterized model

© The Author(s), under exclusive licence to Springer International Publishing AG, part of Springer Nature 2018
M. Braun et al., *Gravitational Biology I*, SpringerBriefs in Space Life Sciences,
https://doi.org/10.1007/978-3-319-93894-3_7

plant continues to provide unprecedented opportunities for isolating mutants and exploring gene function and signaling pathway regulation (Provart et al. 2016).

7.2 Plants Sense Gravity

Plants constantly undergo rhythmic changes in growth such as the diurnal waving of the leaves or the wavy growth of roots (Barlow 2015). They are therefore exposed to constant changes in their relative position to the gravitational field. The environment also imposes changes in the direction of growth such as when, for example, primary root growth is impeded by an impenetrable obstacle in the substrate which requires the root to traverse the surface of the barrier. Such changes are sensed and integrated into overarching developmental programs which control the overall architecture of the plant (Noll 1900; Monshausen and Gilroy 2009). Plants are equipped with exquisitely sensitive sensory machineries which are even able to sense the periodic alterations of the gravitational force caused by movement of the Moon (Barlow and Fisahn 2012; Barlow 2015). This lunar tidal acceleration affects stem elongation growth, leaf movement, seed imbibition and germination (Zajaczkowska and Barlow 2017). Rhythmic variations in the elongation of *Arabidopsis* roots have been described which correspond perfectly to the rise and fall of the gravimetric tide (Fisahn et al. 2012). Plants that were grown on the International Space Station (ISS) and were exposed to lunar gravity in a centrifuge have shown different movement profiles (Fisahn et al. 2015).

7.3 Root Growth Responses

Root growth is heavily influenced by endogenous regulatory rhythms. These oscillatory responses give rise to two growth behaviors: skewing and waving (Roy and Bassham 2014). Obtaining a molecular understanding of skewing and waving has been thought to provide key insights into root growth strategies and in the mechanisms underlying the building of the root's architecture (Braybrook 2017; Roy and Bassham 2017). Oscillatory growth is observed in the laboratory by growing roots on tilted impermeable agar surfaces or in microchips (Grossmann et al. 2011). However, waving is also considered to be part of the roots growth strategy in soil (Tan et al. 2015). In *Arabidopsis*, some but not all ecotypes show skewing behavior, whereas root waving seems to be inherent to all *Arabidopsis* ecotypes. Differentially expressed candidate genes have been identified from different skewing and non-skewing *Arabidopsis* accessions by transcriptome profiling. Genes found to influence skewing or waving using this approach play roles in diverse cellular processes including sugar transport, salt signaling, cell wall organization and hormone signaling (Schultz et al. 2017).

In more than 300 publications, waving and skewing have been considered to be part of the gravitropic response (Roux 2012). However, imaging experiments performed on the ISS in which growth patterns of *Arabidopsis* roots were analyzed in a microgravity environment, clearly demonstrated that waving and skewing were not guided by gravity, but instead by light. These experiments, in which factors such as acceleration vectors, airflow or other directional environmental factors were accounted for, revealed that roots grew away from light whereas their 1-g controls in a centrifuge oriented positively gravitropically in parallel to the acceleration vector (Paul et al. 2012). From this, it is justified to conclude that skewing and waving are driven by plant endogenous rhythms independent of microgravity. Interestingly this phenomenon was reanalyzed using roots from *Medicago trunculata*; here the speed of growth and size of the root enabled an easy nondestructive examination of geometric parameters during root coiling, waving, skewing and gravity mediated curvature (Tan et al. 2015). Data showed that gravity-influenced developmental switches control the root's growth direction and are governed by the root's ability to measure the direction of gravity with some precision. Analysis suggests that, in a similar manner to bacterial chemotaxis, *Medicago* roots are apparently able to find the path of steepest descent by sensing their orientation relative to gravity. This grow-and-switch gravitropism may have provided *Medicago* with an evolutionary favorable trait which allows it to thrive in highly obstructed environments.

7.4   Root Systems Architecture Is Built by Periodic Growth Responses

Lateral roots, organs crucial for exploring the soil, extract nutrients and communicate with the soil microbiome and are initiated post-embryonically in response to environmental cues. Their growth therefore largely defines the complexity of a root system. How then does gravity trigger the periodic growth responses of lateral roots? It was shown in *Arabidopsis thaliana* that lateral root initiation is induced by gravitropic curvature (Ditengou et al. 2008). At the site of lateral root induction auxin accumulated before formation of the primordium, an accumulation which is correlated with a subcellular relocalization of the auxin efflux carrier PIN1 in a single protoxylem cell (Fig. 7.1). This relocalization preceded auxin-dependent gene transcription and defined a competence zone in which lateral root primordia formation became possible (Ditengou et al. 2008).

Root bending bypassed ARF1/19-dependent nuclear auxin signaling. These transcription factors are normally necessary for lateral root formation as shown by analysis of *arf7/19* double knock-out mutants which normally form no lateral roots. However, lateral root formation proceeded upon bending when the root tip was removed (Ditengou et al. 2008). Another feature of lateral root induction is periodic oscillatory changes in auxin levels revealed by either using fluorescent or luminescent auxin reporters. Several studies have identified dynamic alterations of auxin at positions where future founder cells for lateral root primordia will be formed (De Smet et al. 2007; Moreno-Risueno et al. 2010; Xuan et al. 2015). Depending

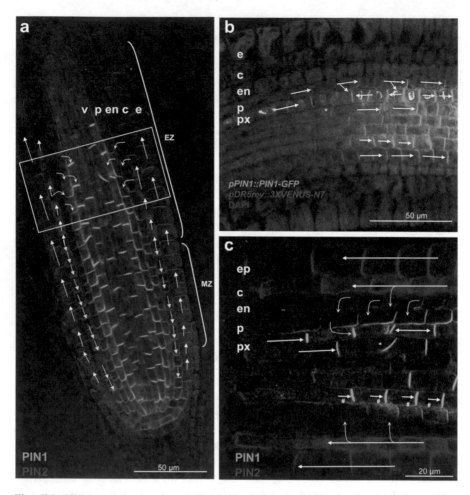

Fig. 7.1 PIN expression and protein localization in gravistimulated roots of wild-type (WT) *Arabidopsis* and p*PIN1::PIN1-GFP*, p*DR5rev::3XVenus-N7* transgenic lines. (**a**) PIN1-PIN2 immunolocalization. Epidermis (e), cortex (c), endodermis (en), pericycle (p), vascular bundle (v), meristematic zone (MZ), elongation zone (EZ), arrows indicate inferred auxin fluxes. (**b**) PIN1-GFP (green) and DR5-Venus (pink), DAPI labeled nuclei (blue), epidermis (e), cortex (c), endodermis (en), pericycle, (p) protoxylem (px), arrows indicate presumed direction of auxin flux, asterisk (*) indicates PIN1 retrieval from basal side of one cell in protoxylem. (**c**) magnified view of boxed area in (A) showing PIN1-PIN2 immunolocalization, epidermis (e), cortex (c), endodermis (en), pericycle (p), vascular bundle (v), arrows indicate presumed direction of auxin flux, hash (#) indicates PIN1 retrieval from lateral side in pericycle cell on the lower side of the root (Ditengou et al. 2008)

on their provenance, the development of these lateral root primordia is more or less sensitive to calcium ions (Richter et al. 2009).

Neither local auxin cues nor their correlation with oscillating auxin maxima can alone explain the patterning of lateral root primordia along the root axis. A recent set of elegant experiments has helped to shed light onto the relationship between the mechanisms behind lateral root initiation and growth (Kircher and Schopfer 2016). Experiments designed to differentiate between bending-induced local auxin accumulation and clock-type oscillations in auxin-induced gene expression revealed that the frequency of lateral roots was promoted by auxin in the mature root; positioning, however, followed a pre-formed pattern determined by previous bending (Kircher and Schopfer 2016).

7.5   Gravitropism Follows Grow-and-Switch or Tipping Point Mechanisms

In the root cap, specialized cells named statocytes sense gravity (cf. Chap. 6). Analysis of gravisensitivity determined a threshold acceleration at between 10^{-3} and 10^{-4} g for roots and 10^{-2} and 10^{-3} g for shoots and a perception time ranging between 1 and 10 s (Perbal et al. 1997; Hejnowicz et al. 1998; Perbal and Driss-Ecole 2003). However, dose-response-based studies, in which it has been suggested that the curvature response varies linearly with the logarithm of the gravity stimulus, may have to be revisited using up-to-date technologies (Perbal et al. 2002).

The weight of an object is the product of gravitational acceleration and its mass. Therefore, all organelles which have a different density than the surrounding cytoplasm and are not fixed in place by cytoskeletal elements may be involved in gravisensing. In *Arabidopsis* roots, relatively dense starch-filled amyloplasts sediment to the bottom of statocytes, where they might stimulate mechanosensors which initiate a chain of physiological events that ends in the reorientation of root growth. In this way, a root which has been turned upside-down regains its previous position. Analysis of starchless phosphoglucomutase loss-of-function (*pgm*) mutants (introduced in Chap. 6) respond to gravity at one-third the rate of wild-type (WT) roots; intermediate mutants (*acg20*, *acg27*) showed responses proportional to their starch content (Kiss et al. 1996). Amyloplasts seem to undergo cage-confined diffusion and cage-breaking motions characteristic of intracellular microenvironments which determine their sedimentation dynamics (Zheng et al. 2015). But which signals and responses are elicited by sedimentation and subsequent mechanical stimulation? While the primary signal is still elusive and no mechanosensitive receptors have been identified to date, interesting candidate genes may be hidden in the glutamate-receptor-like gene family which is homologous to mammalian ionotropic glutamate receptors (Roy et al. 2008; De Bortoli et al. 2016). GLR3.3 for example is such a ligand-gated, Ca^{2+}-permeable channel worth further detailed study as, when mutated, careful dissection of the gravitropic response showed a range of intriguing growth phenotypes (Miller et al. 2010).

7.6   Auxin Is an Early Gravistimulation Signal

'The Power of Movement in Plants', one of the early benchmarks for the study of plant physiology, aimed to understand the principles of tropic growth (Darwin, 1880). This pioneering work underpinned the discovery that the asymmetric distribution of auxin across an organ is a common module for the transmission of an external stimulus into a directional growth response. However, the most pressing task arising from this work still remains: understanding how a gravitropic stimulus leads to an asymmetric auxin distribution. By integrating high-resolution imaging with computer vision-based analysis, several studies have suggested that ROS, pH, and Ca^{2+}-mediated signals all play crucial roles in the early gravity response (Scott and Allen 1999; Fasano et al. 2001; Joo et al. 2001; Hou et al. 2004; Monshausen et al. 2011; Salmi et al. 2011; Hayatsu and Suzuki 2015; Sato et al. 2015; Dummer et al. 2016; Krieger and Shkolnik 2016; Singh et al. 2016; Ponce et al. 2017). A surprising observation reported by Weerasinghe et al. (2009) places the release of ATP from root tip cells downstream of mechanical stimulus-dependent rapid changes in cytosolic Ca^{2+}, but candidate genes playing roles in early gravitropic signaling have not yet been found. Protein kinases are obvious candidates; for example, members of the CrRLK1 receptor-like kinase family are thought to play roles in the earliest stages of strain-activated Ca^{2+}-signaling. Hypotheses which link auxin- and calcium-based signals are strengthened by feedback regulation between the two processes; the external application of auxin modulates gravitropic signaling in a Ca^{2+}-dependent fashion. Application of 100 nM auxin to the root tip triggered a rise in cytosolic Ca^{2+} within 7 sec eliciting a wave of Ca^{2+} which moves shootward back along the root axis (Monshausen et al. 2011).

The pathway resulting in elicitation of a similar wave of shootward-travelling auxin seems now to be well understood. It is widely accepted that the products of two genes: *ARG1* (*ALTERED RESPONSE TO GRAVITY*) and its paralog *ARG-LIKE2* (*ARL-2*) link amyloplast sedimentation with auxin flux (Fukaki et al. 1997; Sedbrook et al. 1999; Guan et al., 2003; Harrison and Masson 2008). ARG1 is a type-II DnaJ-like protein with a C-terminal coil-coiled domain which interacts with Hsp70 molecular chaperones to help disassemble clathrin triskelia (a three-legged pinwheel-shaped heteropolymer, which coats certain post-Golgi and ER vesicles) from clathrin-coated vesicles during endocytosis. Mutations in ARG1 and ARL2 show reduced cytoplasmic alkalinization and a significantly reduced gravitropic response. After combining *arg1* and *arl2* alleles with starch-less mutant *pgm-1*, it was demonstrated that the proteins act in distinct pathways, with PGM affecting the mechanostimulatory response and ARG1 and ARL2 playing roles in the endocytotic modulation of PIN relocalization. Interestingly, AUX1, another polar auxin transport protein (see below), seems not to be affected (Boonsirichai et al. 2003). Using a rotating stage image system, Evans and coworkers were able to maintain the root tip at a constant angle and observe at constant rate the gravitropic response regardless of the angle of tip orientation (Mullen et al. 2000; Wolverton et al. 2002). WT and *pin3–1* plants showed, in contrast to *pgm-1*, increasing response rates as the tip was constrained at greater angles. Interestingly, wild type roots expressing the auxin response reporter DR5::GFP displayed a graded GFP response with a maximum along the lower flank of wild type roots, whereas *pgm-1* roots formed a GFP maximum in the central columella but lacked any observable gradient at up to 6 h after reorientation (Wolverton et al. 2011). Based on the quantitative analysis of plastid sedimentation, a relationship between root cap angle and gravitropic response was found.

These data are consistent with the idea of several overlapping sensory response networks involved in controlling gravitropism, with PIN3 performing a rate-limiting early response but becoming less important for sustained differential growth (Friml et al. 2002; Wolverton et al. 2011). Rapid gravity-stimulated changes in PIN3 polarity have been observed; however, PIN3 polarization apparently does not require secretion of *de novo* synthesized proteins or protein degradation but rather uses (probably ARG1/ARL2-dependent) clathrin-dependent endocytosis for rapid cellular relocalization. This pathway, which distributes auxin asymmetrically during the response to gravity, has been suggested to require Brefeldin A-sensitive recycling and recruitment of an ARF-GEF (guanine nucleotide exchange factor for ARF GTPases) for polar targeting and rapid transcytotic relocation to different sides of graviresponsive cells (Steinmann et al. 1999; Friml et al. 2002; Kleine-Vehn et al. 2010; Naramoto et al. 2010).

Flux of auxin through plant organs, tissues and cells not only requires efflux from cells but also influx and intracellular fluxes as recently demonstrated (Middleton et al. 2018). The first auxin influx carrier prototype was cloned from the agravitropic *aux1* mutant (Bennett et al. 1996). Over the years genetic and functional analysis revealed important aspects of its function. Surprisingly AUX1 is not only a major auxin import carrier for auxin (indole-3-acetic acid) uptake but also mediates membrane depolarization and correlates with a long-distance Ca^{2+} wave that modulates the auxin response (Dindas et al. 2018). This effect has found additional support from studies on root hairs, in which the external or internal application of auxin caused Ca^{2+} changes (observed with R-GECO1) which were propagated as long-distance waves. These IAA-triggered local and systemic calcium signals were blocked by treatment with the SCFTIR1/AFB 46 -signaling inhibitor 47 auxinole and appeared strongly impaired in the *tir1afb2afb3* triple mutant (Dindas et al. 2018). It can be hypothesized that AUX1 operates as the major auxin re-uptake route after auxin is released from cells by different PIN efflux carriers in the gravisensing region of the root cap. Ca^{2+} waves may also well operate in this region just as they do in root hairs.

7.7   Gravitropic Signaling by Ca²⁺/Calmodulin-Dependent Kinase CRK5

It is therefore not surprising that a Ca^{2+}/calmodulin-dependent kinase, CRK5, has been found in this central gravisensing zone (Rigo et al. 2013). CRK5 belongs to an eight-member CRK family in *Arabidopsis*, and shares structural and functional features with CRKs found in other plants (Harper et al. 2004). Its plasma-membrane localization is probably mediated by a myristoylation/plamitoylation modification and is polar in statocytes, forming a U-shaped zone towards the root tip (Figs. 7.2 and 7.3). Absence of CRK5 in a T-DNA mutant did not affect the localization of AUX1 or PIN1, PIN3, PIN4 or PIN7 in the root cap and gravisensing statocytes. However, a PIN2-GFP reporter was depleted in the transition zone between root cap and elongation zones. There was also a remarkable basal-to-apical shift of

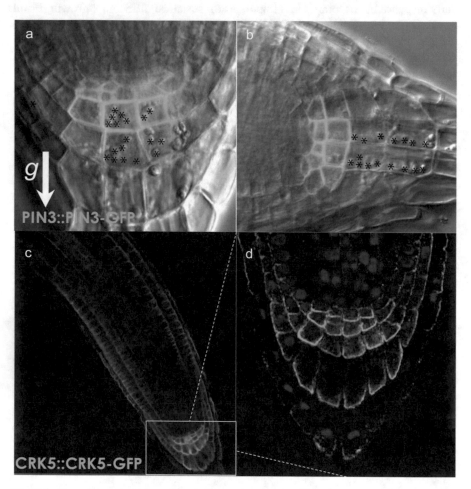

Fig. 7.2 Expression of PIN3 and CRK5 in columella cells. Six days old Arabidopsis thaliana seedlings expressing respectively, (**a–b**) and CRK5::CRK5-GFP (**c-d**). (**a**) PIN3::PIN3-GFP in vertically grown plants. In columella cells PIN3-GFP localizes all around the cell whereas statoliths (*) are distributed at the bottom of the cells. (**b**) PIN3::PIN3-GFP in gravistimulated (90°) roots for 1 h. Statoliths and PIN3 re-localize laterally in columella cells. (**c**) CRK5::CRK5-GFP localization at the root tip. (**d**) Enlargement of the boxed area in (**c**) showing nuclei (blue) and the polar localization of CRK5-GFP in columella and lateral root cap cells. White arrows in (**a**) and (**c**) indicate the direction of the gravity vector

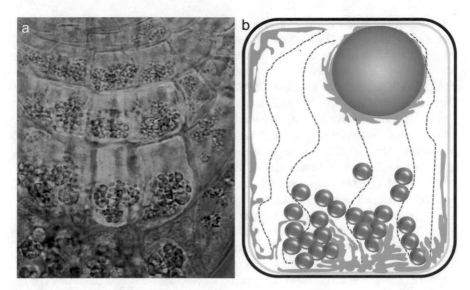

Fig. 7.3 The gravitropic signaling network. (**a**) Arabidopsis root expressing ER-localized auxin marker DR5::revGFP (green) (Ottenschlager et al. 2003). Nuclei (blue) stained with DAPI. (**b**) Schematic representation of a columella cell displaying gravitropic signaling components such as actin cytoskeleton (blue dashed line), ER network (green), and statoliths (dark grey)

PIN2-GFP which potentially caused the observed significant delay to the gravitropic response, as in a *crk-5* null mutation, both positive and negative gravitropic bending of roots and shoots was inhibited (Rigo et al. 2013). A corresponding reduction in the auxin response along the lower side of roots after gravistimulation suggested a lower auxin content there and hence a reduced rate of auxin flux from statocytes into the epidermal cell layers. Such changes are able to account for the observed 30% reduction of root growth rate compared to the wild type. There is evidence for functional overlap with a second protein family important to the gravitropic response, which is also regulated by calcium ions. Alongside the AGC kinases PID, WAG1, WAG2, MAPK and D6PKs, CRKs probably regulate cell-type specific phosphorylation of specific residues in hydrophilic loops of certain PINs to regulate their polar membrane recycling (Ganguly and Cho 2012; Ditengou et al. 2018; Dory et al. 2018). So far the position of the different members of the CRK family in the gravitropic signaling pathway is not clear, but they may play important roles in regulating the activities of activity and membrane localization of other PIN-specific AGC kinases (Weller et al. 2017).

7.8 Downstream Regulation of PIN Function

Auxin fluxes are regulated on several levels. The PIN efflux carriers constitute the basic machinery capable of releasing negatively charged indole-3-acetic acid anions from the cytoplasm into the apoplast. The rate and direction of this release is regulated by multiple kinases. Although it has previously been proposed that specific serine residues are modified by specific kinases either to control auxin flux or PIN localization (Michniewicz et al. 2007; Zourelidou et al. 2014), recent studies using phospho-specific antibodies show that the same sites are targeted by kinases which regulate each process (for example D6PK and PINOID), meaning the regulation is likely to be complex, dynamic and context-specific (Weller et al. 2017).

Other regulatory mechanisms are likely also to be at work. For example, it has been known for decades that the localized synthesis and directed transport of flavonols, plant specific phenolic compounds, modulate auxin transport and gravitropism (Buer and Muday 2004; Buer et al. 2007). Experiments with the synthetic polar auxin transport inhibitor (and functional flavonol analog) N-1-naphthylphthalamic acid (NPA) and *transparent testa* mutants with altered flavonol levels revealed altered gravitropic responses (Taylor and Grotewold 2005; Teale and Palme 2018). Hence application of nanomolar levels of the flavonol quercetin to the *pin2* mutants with strong defects in gravitropic response was sufficient to restore wild type-like auxin distribution patterns and partially restore the gravitropic response (Santelia et al. 2008). The question of whether NPA and flavonols can be considered to be functionally equivalent with respect to the gravity response is fairly complex and considered elsewhere (Teale and Palme 2018). However, observations may suggest that flavonoids possibly exert their function by directly binding to PIN proteins and regulating their auxin transport capacity and subcellular localization (Buer and Muday 2004; Kuhn et al. 2017).

Other proteins possibly involved in polar auxin transport and gravitropic response are ATP-binding cassette (ABC) transporter family proteins (Nagashima et al. 2008) with the anion channel blocker 5-nitro-2-(3-phenylpropylamino)-benzoic acid inhibited ABC-dependent transport activity in a heterologous assay and the root gravitropic response (Cho et al. 2014). Mechanistically it has been suggested that members of this family may directly form complexes with PIN proteins on the protein level, but solid evidence is yet lacking to back up this hypothesis.

7.9   Interaction Between Auxin and Other Hormones

Several lines of evidence suggest auxin works in concert with other plant hormones to regulate root gravitropism (Philosoph-Hadas et al. 2005). For example, gibberellic acid (GA) shows asymmetric action during gravitropic responses (Löfke et al. 2013). GA signaling at the lower side of the root stabilizes the auxin transporter PIN-FORMED2 (PIN2) at the plasma membrane through a specific GA effect on protein trafficking lytic vacuoles, hence suggesting an interplay between asymmetric auxin and gibberellin activities in the modulation of auxin fluxes for root gravitropic responses.

Besides GA, it was shown that cytokinin synthesized in root cap cells redistributes towards the lower side of the gravistimulated root within minutes, suggesting that cytokinin is acting as early during root gravitropism as well as auxin (Aloni et al.

2004, 2006). This suspicion was reinforced by the fact that exogenous cytokinin application to vertical roots induced root bending towards the application site (Aloni et al. 2004).

Ethylene also interacts with auxin to regulate root gravitropism. This relationship has been firmly established from the discovery that the agravitropic Arabidopsis *pin2* mutant was allelic with the ethylene insensitive mutants *ein1* and *agr1* (Müller et al. 1998). This interaction is complicated and multi-faceted, affecting not only auxin efflux but the biosynthesis and signaling of both hormones. Furthermore, although the agravitropic phenotype of *pin2/agr1/ein1* is specific to the root, the overall influence of the auxin/ethylene interaction is broad, not only affecting gravitropism, but also regulating other important processes which respond to a mixture of external and internal stimuli such as bending of the hypocotyl hook in etiolated seedlings, root elongation and root hair development (Stepanova and Alonso 2005; Stepanova et al. 2007). The regulation of auxin biosynthesis by ethylene is an aspect of the relationship between the two hormones which has also been difficult to parse, with ethylene inducing the expression of auxin biosynthetic genes, such as WEAK ETHYLENE INSENSITIVE 2 (WEI2)/ANTHRANILATE SYNTHASE α1 (ASA1), WEI7/ASB1, WEI8/TRYPTOPHAN AMINOTRANSFERASE OF ARABIDOPSIS 1 (TAA1)/TRANSPORT INHIBITOR RESPONSE2 (TIR2) and its homolog TAR1 (Stepanova et al. 2005, 2008). Ethylene promotes auxin transport in the root in both directions by upregulating the expression of several transcripts encoding auxin transporters, including *PIN1*, *PIN2*, *PIN4* and *AUX1* (Ruzicka et al. 2007; Negi et al. 2008; Vandenbussche et al. 2010; Lewis et al., 2011; Muday et al. 2012).

Ethylene is synthesized from its precursor 1-aminocyclopropane-1-carboxylic acid (ACC), which is, in turn, synthesized by ACETYL-COA SYNTHETASES (ACS). Auxin induces the expression of *ACS* genes, increasing the production of ethylene (Woeste et al. 1999; Tsuchisaka and Theologis 2004). Crosstalk between auxin and ethylene has been most clearly demonstrated in a series of experiments which showed that by removing specific auxin and ethylene responses by using *aux1* and *ein2* genotypes respectively, underlying cross-regulation by ethylene and auxin could be observed. (Stepanova et al. 2007). A picture is emerging whereby auxin mutants also lose sensitivity to ethylene, but ethylene mutants retain their sensitivity to auxin. Such observations suggest that ethylene signaling acts through auxin and not *vice versa* (Vandenbussche et al. 2012). This conclusion appears to be borne out as the EIN3-dependent ethylene response in the root transition zone requires high auxin activity (Stepanova et al. 2007), as accumulation of the protein is enhanced by auxin. The auxin-mediated repression of two F-box proteins, EBF1 and EBF2 has been implicated here as both proteins mediate EIN3 degradation (He et al. 2011).

The synergistic impact of ethylene and auxin on the asymmetric growth of the gravitropic root also involves cytokinin. Street and colleagues (Street et al. 2016) reported that the aforementioned cytokinin regulation of root cell elongation occurs through ethylene-dependent and -independent mechanisms, both hormonal signals converging on AUX1 as a regulatory hub (Street et al. 2016).

7.10   The *Arabidopsis* Transcriptome Is Affected by Altered Gravity

We still do not fully understand how gravity affects the expression of gravity-sensitive genes and whether particular molecular signatures are elicited by changing the orientation of roots in the gravitational field. In particular it will be important to clarify which genes are necessary to characterize different phases of the gravity response. Also important is the question of whether changing gravity causes a specific stress response against which cells may remodel their metabolic pathways in order to compensate. For this, it will be crucial to identify the genes which regulate the gravity response. The analysis of the gravitome, the complement of gravity regulated genes, is likely to improve our understanding of the molecular mechanisms regulating perception, transduction and the response to gravity and how this response is regulated (Aubry-Hivet et al. 2014). Although microgravity can be simulated in ground-based facilities by averaging gravity to zero levels, here the influence of gravity will never fully be neutralized (Briegleb 1992; Herranz et al. 2013a, b). Therefore, instead, experiments may only be performed on appropriate space environment platforms such as on the International Space Station (ISS), satellites, sounding rockets, drop towers, or aircrafts during parabolic flight (cf Chap. 2). Parabolic flights are more easily accessible for experimentation and offer experimental scenarios that enable us to obtain independent experimental replications for solid statistical evaluation which are typically elusive in other space flight scenarios. Although in these experiments, microgravity phases are followed by phases of hyper-g accelerations, carefully designed control experiments have been able to separate the effects of microgravity and hyper-g on the biological samples (Paul et al. 2011; Herranz et al. 2013a, b; Aubry-Hivet et al. 2014; Herranz and Medina 2014). By using mutants of the auxin efflux pathway (i.e. *pin2*, *pin3*) it has been possible to correlate, under parabolic flight conditions, genetic relationships with remodeling of metabolic pathways. In roots lacking columella-localized PIN3, changes in gene expression were more dramatic than in those defective in the epidermis and cortex cell-specific PIN2 confirming a critical function of PIN3 in mediating auxin-fluxes in response to altered gravity (Aubry-Hivet et al. 2014). These and other studies have provided us with many of the important insights which are needed if we are to understand signal transduction processes in altered gravity conditions. Similar transcriptome studies performed on *Arabidopsis* seedlings or *in vitro* grown callus cultures which were exposed to microgravity either during growth in suborbital flights or on the ISS provided further evidence that space provides an environment which requires novel adaptive processes (Paul et al. 2011). Interestingly comparison of knockout mutants of *ARG1* with wild type *Arabidopsis* revealed the engagement of unique genes during physiological adaptation to the space flight environment.

7.11   High-Resolution Imaging of Plant Cells in Altered Gravity

Transcriptomic analyses performed over the years have generated a tremendous amount of data about genes differentially regulated under various gravity conditions. However, it is not known in which tissues and cells these genes are expressed and whether resulting proteins dynamically respond to changes in gravity vector. These questions are beginning to find answers thanks to new advanced microscopy techniques and protocols. A correlation between gene expression analysis and protein subcellular localization was shown recently in mammalian FTC-133 cancer cells expressing the LifeAct-GFP marker protein for the visualization of F-actin, using the compact fluorescence microscope (FLUMIAS) for fast live-cell imaging under real microgravity as provided by a parabolic flight and sounding rocket (Corydon et al. 2016). More specific tools such as the iRoCS (intrinsic root coordinate system), which now enables direct and quantitative comparison between the root tips of plant populations at single-cell resolution (Schmidt et al. 2014), are needed and will for sure allow to combine omics with imaging data.

7.12   Outlook

Space biology offers unique platforms for plant biology to study the molecular mechanisms of adaptive behavior in plants. According to the roadmap of several space agencies, plants will be an integrated part of human space exploration, providing biologically based life support for food, water recycling and health. The EU funded CEADSE (Controlled Environment Agriculture Development for Space and Earth) project for instance has developed interesting hardware for future space-based bioregenerative life support systems which may in future not only provide a useful prototype platform for plant growth in space on the ISS but also other extreme environments like Antarctica (https://cordis.europa.eu/result/rcn/182930_en.html).

However, as well as plants helping us to explore the mysteries of space, space travel is uniquely placed to help us explore the mysteries of plants. The influence of gravity over plant stature is pervasive and difficult to understand in experimental environments in which it cannot be removed. If we are able to understand the molecular events which shape plant stature in the hope that we may one day be able to design platforms more suited to highly artificial modern agricultural settings, we must first fully understand them. Microgravity experiments have, in this context, the potential to be used as an exceptionally useful tool for the fine dissection of sensitive pathways which, on Earth, lie hidden.

References

Aloni R, Langhans M, Aloni E, Ullrich CI (2004) Role of cytokinin in the regulation of root gravitropism. Planta 220:177–182

Aloni R, Aloni E, Langhans M, Ullrich CI (2006) Role of Cytokinin and Auxin in shaping root architecture: regulating vascular differentiation, lateral root initiation, root apical dominance and root Gravitropism. Ann Bot 97:883–893

Aubry-Hivet D, Nziengui H, Rapp K, Oliveira O, Paponov IA, Li Y, Hauslage J, Vagt N, Braun M, Ditengou FA, Dovzhenko A, Palme K (2014) Analysis of gene expression during parabolic flights reveals distinct early gravity responses in Arabidopsis roots. Plant Biol 16:129–141

Barlow PW (2015) Leaf movements and their relationship with the lunisolar gravitational force. Ann Bot 116:149–187

Barlow PW, Fisahn J (2012) Lunisolar tidal force and the growth of plant roots, and some other of its effects on plant movements. Ann Bot 110:301–318

Bennett MJ, Marchant A, Green HG, May ST, Ward SP, Millner PA, Walker AR, Schulz B, Feldmann KA (1996) Arabidopsis AUX1 gene: a permease-like regulator of root gravitropism. Science 273:948–950

Boonsirichai K, Sedbrook JC, Chen RJ, Gilroy S, Masson PH (2003) ALTERED RESPONSE TO GRAVITY is a peripheral membrane protein that modulates gravity-induced cytoplasmic alkalinization and lateral auxin transport in plant statocytes. Plant Cell 15:2612–2625

Braybrook SA (2017) Plant development: lessons from getting it twisted. Curr Biol 27:R758–R760

Briegleb W (1992) Some qualitative and quantitative aspects of the fast-rotating clinostat as a research tool. ASGSB Bull 5:23–30

Buer CS, Muday GK (2004) The transparent testa4 mutation prevents flavonoid synthesis and alters auxin transport and the response of Arabidopsis roots to gravity and light. Plant Cell 16:1191–1205

Buer CS, Muday GK, Djordjevic MA (2007) Flavonoids are differentially taken up and transported long distances in Arabidopsis. Plant Physiol 145:478–490

Cho M, Henry EM, Lewis DR, Wu GS, Muday GK, Spalding EP (2014) Block of ATP-binding cassette B19 ion channel activity by 5-Nitro-2-(3-Phenylpropylamino)-benzoic acid impairs polar auxin transport and toot gravitropism. Plant Physiol 166:2091–2099

Corydon TJ, Kopp S, Wehland M, Braun M, Schutte A, Mayer T, Hulsing T, Oltmann H, Schmitz B, Hemmersbach R, Grimm D (2016) Alterations of the cytoskeleton in human cells in space proved by life-cell imaging. Sci Rep 6:20043

Darwin C (1880) The power of movement in plants. John Murray, London

De Bortoli S, Teardo E, Szabò I, Morosinotto T, Alboresi A (2016) Evolutionary insight into the ionotropic glutamate receptor superfamily of photosynthetic organisms. Biophys Chem 218:14–26

De Smet I, Tetsumura T, De Rybel B, Frey NFD, Laplaze L, Casimiro I, Swarup R, Naudts M, Vanneste S, Audenaert D, Inze D, Bennett MJ, Beeckman T (2007) Auxin-dependent regulation of lateral root positioning in the basal meristem of Arabidopsis. Development 134:681–690

Dindas J, Scherzer S, Roelfsema MRG, von Meyer K, Muller HM, Al-Rasheid KAS, Palme K, Dietrich P, Becker D, Bennett MJ, Hedrich R (2018) AUX1-mediated root hair auxin influx governs SCF(TIR1/AFB)-type Ca(2+) signaling. Nat Commun 9:1174

Ditengou FA, Teale WD, Kochersperger P, Flittner KA, Kneuper I, Van Der Graaff E, Nziengui H, Pinosa F, Li X, Nitschke R, Laux T, Palme K (2008) Mechanical induction of lateral root initiation in Arabidopsis thaliana. Proc Natl Acad Sci U S A 105:18818–18823

Ditengou FA, Gomes D, Nziengui H, Kochersperger P, Lasok H, Medeiros V, Paponov IA, Nagy SK, Nadai TV, Meszaros T, Barnabas B, Ditengou BI, Rapp K, Qi LL, Li XG, Becker C, Li CY, Doczi R, Palme K (2018) Characterization of auxin transporter PIN6 plasma membrane targeting reveals a function for PIN6 in plant bolting. New Phytol 217:1610–1624

Dory M, Hatzimasoura E, Kallai BM, Nagy SK, Jager K, Darula Z, Nadai TV, Meszaros T, Lopez-Juez E, Barnabas B, Palme K, Bogre L, Ditengou FA, Doczi R (2018) Coevolving MAPK and

PID phosphosites indicate an ancient environmental control of PIN auxin transporters in land plants. FEBS Lett 592:89–102

Dummer M, Michalski C, Essen LO, Rath M, Galland P, Forreiter C (2016) EHB1 and AGD12, two calcium-dependent proteins affect gravitropism antagonistically in Arabidopsis thaliana. J Plant Physiol 206:114–124

Fasano JM, Swanson SJ, Blancaflor EB, Dowd PE, Kao TH, Gilroy S (2001) Changes in root cap pH are required for the gravity response of the Arabidopsis root. Plant Cell 13:907–921

Fisahn J, Yazdanbakhsh N, Klingele E, Barlow P (2012) Arabidopsis thaliana root growth kinetics and lunisolar tidal acceleration. New Phytol 195:346–355

Fisahn J, Klingele E, Barlow P (2015) Lunar gravity affects leaf movement of Arabidopsis thaliana in the international Space Station. Planta 241:1509–1518

Friml J, Wisniewska J, Benkova E, Mendgen K, Palme K (2002) Lateral relocation of auxin efflux regulator PIN3 mediates tropism in Arabidopsis. Nature 415:806–809

Fukaki H, Fujisawa H, Tasaka M (1997) The RHG gene is involved in root and hypocotyl gravitropism in Arabidopsis thaliana. Plant Cell Physiol 38:804–810

Ganguly A, Cho H-T (2012) The phosphorylation code is implicated in cell type-specific trafficking of PIN-FORMEDs. Plant Signal Behav 7:1215–1218

Grossmann G, Guo WJ, Ehrhardt DW, Frommer WB, Sit RV, Quake SR, Meier M (2011) The RootChip: an integrated microfluidic chip for plant science. Plant Cell 23:4234–4240

Guan CH, Rosen ES, Boonsirichai K, Poff KL, Masson PH (2003) The ARG1-LIKE2 gene of Arabidopsis functions in a gravity signal transduction pathway that is genetically distinct from the PGM pathway. Plant Physiol 133:100–112

Harper JE, Breton G, Harmon A (2004) Decoding Ca2+ signals through plant protein kinases. Annu Rev Plant Biol 55:263–288

Harrison BR, Masson PH (2008) ARL2, ARG1 and PIN3 define a gravity signal transduction pathway in root statocytes. Plant J 53:380–392

Hayatsu M, Suzuki S (2015) Electron probe X-ray microanalysis studies on the distribution change of intra- and extracellular calcium in the elongation zone of horizontally reoriented soybean roots. Microscopy (Oxf) 64:327–334

He W, Brumos J, Li H, Ji Y, Ke M, Gong X, Zeng Q, Li W, Zhang X, An F, Wen X, Li P, Chu J, Sun X, Yan C, Yan N, Xie DY, Raikhel N, Yang Z, Stepanova AN, Alonso JM, Guo H (2011) A small-molecule screen identifies L-kynurenine as a competitive inhibitor of TAA1/TAR activity in ethylene-directed auxin biosynthesis and root growth in Arabidopsis. Plant Cell 23:3944–3960

Hejnowicz Z, Sondag C, Alt W, Sievers A (1998) Temporal course of graviperception in intermittently stimulated cress roots. Plant Cell Environ 21:1293–1300

Herranz R, Medina FJ (2014) Cell proliferation and plant development under novel altered gravity environments. Plant Biol 16:23–30

Herranz R, Anken R, Boonstra J, Braun M, Christianen PCM, de Geest M, Hauslage J, Hilbig R, Hill JA, Lebert M, Medina J, Vagt N, Ullrich O, van JWA L, Hemmersbach R (2013a) Ground-based facilities for simulation of microgravity, including terminology and organism-specific recommendations for their use. Astrobiology 13. https://doi.org/10.1089/ast.2012.0876

Herranz R, Anken R, Boonstra J, Braun M, Christianen PCM, de Geest M, Hauslage J, Hilbig R, Hill RJA, Lebert M, Medina FJ, Vagt N, Ullrich O, van Loon JJWA, Hemmersbach R (2013b) Ground-based facilities for simulation of microgravity: organism-specific recommendations for their use, and recommended terminology. Astrobiology 13:1–17

Hou G, Kramer VL, Wang YS, Chen R, Perbal G, Gilroy S, Blancaflor EB (2004) The promotion of gravitropism in Arabidopsis roots upon actin disruption is coupled with the extended alkalinization of the columella cytoplasm and a persistent lateral auxin gradient. Plant J 39:113–125

Joo JH, Bae YS, Lee JS (2001) Role of auxin-induced reactive oxygen species in root gravitropism. Plant Physiol 126:1055–1060

Kircher S, Schopfer P (2016) Priming and positioning of lateral roots in Arabidopsis. An approach for an integrating concept. J Exp Bot 67:1411–1420

Kiss JZ, Wright JB, Caspar T (1996) Gravitropism in roots of intermediate-starch mutants of Arabidopsis. Physiol Plant 97:237–244

Kleine-Vehn J, Ding Z, Jones AR, Tasaka M, Morita MT, Friml J (2010) Gravity-induced PIN transcytosis for polarization of auxin fluxes in gravity-sensing root cells. Proc Natl Acad Sci U S A 107:22344–22349

Krieger G, Shkolnik D (2016) Reactive oxygen species tune root tropic responses. Plant Physiol 172:1209–1220

Kuhn BM, Nodzynski T, Errafi S, Bucher R, Gupta S, Aryal B, Dobrev P, Bigler L, Geisler M, Zazimalova E, Friml J, Ringli C (2017) Flavonol-induced changes in PIN2 polarity and auxin transport in the Arabidopsis thaliana roll-2 mutant require phosphatase activity. Sci Rep 7:41906

Lewis DR, Negi S, Sukumar P, Muday GK (2011) Ethylene inhibits lateral root development, increases IAA transport and expression of PIN3 and PIN7 auxin efflux carriers. Development 138:3485–3495

Löfke C, Zwiewka M, Heilmann I, Van Montagu MCE, Teichmann T, Friml J (2013) Asymmetric gibberellin signaling regulates vacuolar trafficking of PIN auxin transporters during root gravitropism. Proc Natl Acad Sci U S A 110:3627–3632

Michniewicz M, Zago MK, Abas L, Weijers D, Schweighofer A, Meskiene I, Heisler MG, Ohno C, Zhang J, Huang F, Schwab R, Weigel D, Meyerowitz EM, Luschnig C, Offringa R, Friml J (2007) Antagonistic regulation of PIN phosphorylation by PP2A and PINOID directs auxin flux. Cell 130:1044–1056

Middleton AM, Dal Bosco C, Chlap P, Bensch R, Harz H, Ren F, Bergmann S, Wend S, Weber W, Hayashi KI, Zurbriggen MD, Uhl R, Ronneberger O, Palme K, Fleck C, Dovzhenko A (2018) Data-driven modeling of intracellular auxin fluxes indicates a dominant role of the ER in controlling nuclear auxin uptake. Cell Rep 22:3044–3057

Miller ND, Brooks TLD, Assadi AH, Spalding EP (2010) Detection of a Gravitropism phenotype in glutamate receptor-like 3.3 mutants of Arabidopsis thaliana using machine vision and computation. Genetics 186:585–U206

Monshausen GB, Gilroy S (2009) The exploring root - root growth responses to local environmental conditions. Curr Opin Plant Biol 12:766–772

Monshausen GB, Miller ND, Murphy AS, Gilroy S (2011) Dynamics of auxin-dependent Ca^{2+} and pH signaling in root growth revealed by integrating high-resolution imaging with automated computer vision-based analysis. Plant J 65:309–318

Moreno-Risueno MA, Van Norman JM, Moreno A, Zhang J, Ahnert SE, Benfey PN (2010) Oscillating gene expression determines competence for periodic Arabidopsis root bBranching. Science 329:1306–1311

Muday GK, Rahman A, Binder BM (2012) Auxin and ethylene: collaborators or competitors? Trends Plant Sci 17:181–195

Mullen JL, Wolverton C, Ishikawa H, Evans ML (2000) Kinetics of constant gravitropic stimulus responses in Arabidopsis roots using a feedback system. Plant Physiol 123:665–670

Müller A, Guan C, Gälweiler L, Tänzler P, Huijser P, Marchant A, Parry G, Bennett M, Wisman E, Palme K (1998) AtPIN2 defines a locus of Arabidopsis for root gravitropism. EMBO J 17:101–109

Nagashima A, Uehara Y, Sakai T (2008) The ABC subfamily B auxin transporter AtABCB19 is involved in the inhibitory effects of N-1-naphthyphthalamic acid on the phototropic and gravitropic responses of Arabidopsis hypocotyls. Plant Cell Physiol 49:1250–1255

Naramoto S, Kleine-Vehn J, Robert S, Fujimoto M, Dainobu T, Paciorek T, Ueda T, Nakano A, Van Montagu MCE, Fukuda H, Friml J (2010) ADP-ribosylation factor machinery mediates endocytosis in plant cells. Proc Natl Acad Sci U S A 107:21890–21895

Negi S, Ivanchenko MG, Muday GK (2008) Ethylene regulates lateral root formation and auxin transport in Arabidopsis thaliana. Plant J 55:175–187

Noll F (1900) Über den bestimmenden Einfluss von Wurzelkrümmungen auf Entstehung und Anordnung der Seitenwurze. Landwirtschaftliche Jahrbucher 29:361–426

Ottenschlager I, Wolff P, Wolverton C, Bhalerao RP, Sandberg G, Ishikawa H, Evans M, Palme K (2003) Gravity-regulated differential auxin transport from columella to lateral root cap cells. Proc Natl Acad Sci U S A 100:2987–2991

Paul A-L, Manak MS, Mayfield JD, Reyes MF, Gurley WB, Ferl RJ (2011) Parabolic flight iInduces changes in gene expression patterns in Arabidopsis thaliana. Astrobiology 11:743–758

Paul AL, Amalfitano CE, Ferl RJ (2012) Plant growth strategies are remodeled by spaceflight. BMC Plant Biol 12:232

Perbal G, Driss-Ecole D (2003) Mechanotransduction in gravisensing cells. Trends Plant Sci 8:498–504

Perbal G, Driss-Ecole D, Tewinkel M, Volkmann D (1997) Statocyte polarity and gravisensitivity in seedling roots grown in microgravity. Planta 203:S57–S62

Perbal G, Jeune B, Lefranc A, Carnero-Diaz E, Driss-Ecole D (2002) The dose-response curve of the gravitropic reaction: a re-analysis. Physiol Plant 114:336–342

Philosoph-Hadas S, Friedman H, Meir S (2005) Gravitropic bending and plant hormones. Plant Horm 72:31–78

Ponce G, Corkidi G, Eapen D, Lledias F, Cardenas L, Cassab G (2017) Root hydrotropism and thigmotropism in Arabidopsis thaliana are differentially controlled by redox status. Plant Signal Behav 12:e1305536

Provart NJ, Alonso J, Assmann SM, Bergmann D, Brady SM, Brkljacic J, Browse J, Chapple C, Colot V, Cutler S, Dangl J, Ehrhardt D, Friesner JD, Frommer WB, Grotewold E, Meyerowitz E, Nemhauser J, Nordborg M, Pikaard C, Shanklin J, Somerville C, Stitt M, Torii KU, Waese J, Wagner D, McCourt P (2016) 50 years of Arabidopsis research: highlights and future directions. New Phytol 209:921–944

Richter GL, Monshausen GB, Krol A, Gilroy S (2009) Mechanical Stimuli Modulate Lateral Root Organogenesis. Plant Physiol 151:1855–1866

Rigo G, Ayaydin F, Tietz O, Zsigmond L, Kovacs H, Pay A, Salchert K, Darula Z, Medzihradszky KF, Szabados L, Palme K, Koncz C, Cseplo A (2013) Inactivation of plasma membrane-localized CDPK-RELATED KINASE5 decelerates PIN2 exocytosis and root gravitropic response in Arabidopsis. Plant Cell 25:1592–1608

Roux SJ (2012) Root waving and skewing - unexpectedly in micro-g. BMC Plant Biol 12:231

Roy R, Bassham DC (2014) Root growth movements: waving and skewing. Plant Sci 221:42–47

Roy R, Bassham DC (2017) TNO1, a TGN-localized SNARE-interacting protein, modulates root skewing in Arabidopsis thaliana. BMC Plant Biol 17:73

Roy SJ, Gilliham M, Berger B, Essah PA, Cheffings C, Miller AJ, Davenport RJ, Liu LH, Skynner MJ, Davies JM, Richardson P, Leigh RA, Tester M (2008) Investigating glutamate receptor-like gene co-expression in Arabidopsis thaliana. Plant Cell Environ 31:861–871

Ruzicka K, Ljung K, Vanneste S, Podhorska R, Beeckman T, Friml J, Benkova E (2007) Ethylene regulates root growth through effects on auxin biosynthesis and transport-dependent auxin distribution. Plant Cell 19:2197–2212

Salmi ML, ul Haque A, Bushart TJ, Stout SC, Roux SJ, Porterfield DM (2011) Changes in gravity rapidly alter the magnitude and direction of a cellular calcium current. Planta 233:911–920

Santelia D, Henrichs S, Vincenzetti V, Sauer M, Bigler L, Klein M, Bailly A, Lee Y, Friml J, Geisler M, Martinoia E (2008) Flavonoids redirect PIN-mediated polar auxin fluxes during root gravitropic responses. J Biol Chem 283:31218–31226

Sato EM, Hijazi H, Bennett MJ, Vissenberg K, Swarup R (2015) New insights into root gravitropic signalling. J Exp Bot 66:2155–2165

Schmidt T, Pasternak T, Liu K, Blein T, Aubry-Hivet D, Dovzhenko A, Duerr J, Teale W, Ditengou FA, Burkhardt H, Ronneberger O, Palme K (2014) The iRoCS toolbox - 3D analysis of the plant root apical meristem at cellular resolution. Plant J 77:806–814

Schultz ER, Zupanska AK, Sng NJ, Paul AL, Ferl RJ (2017) Skewing in Arabidopsis roots involves disparate environmental signaling pathways. BMC Plant Biol 17:31

Scott AC, Allen NS (1999) Changes in cytosolic pH within Arabidopsis root columella cells play a key role in the early signaling pathway for root gravitropism. Plant Physiol 121:1291–1298

Sedbrook JC, Chen R, Masson PH (1999) ARG1 (altered response to gravity) encodes a DnaJ-like protein that potentially interacts with the cytoskeleton. Proc Natl Acad Sci U S A 96:1140–1145

Singh R, Singh S, Parihar P, Mishra RK, Tripathi DK, Singh VP, Chauhan DK, Prasad SM (2016) Reactive oxygen species (ROS): beneficial companions of plants' developmental processes. Front Plant Sci 7:1299

Steinmann T, Geldner N, Grebe M, Mangold S, Jackson CL, Paris S, Galweiler L, Palme K, Jurgens G (1999) Coordinated polar localization of auxin efflux carrier PIN1 by GNOM ARF GEF. Science 286:316–318

Stepanova AN, Alonso JM (2005) Arabidopsis ethylene signaling pathway. Sci STKE 2005:1–4

Stepanova AN, Hoyt JM, Hamilton AA, Alonso JM (2005) A link between ethylene and auxin uncovered by the characterization of two root-specific ethylene-insensitive mutants in Arabidopsis. Plant Cell 17:2230–2242

Stepanova AN, Yun J, Likhacheva AV, Alonso JM (2007) Multilevel interactions between ethylene and auxin in Arabidopsis roots. Plant Cell 19:2169–2185

Stepanova AN, Robertson-Hoyt J, Yun J, Benavente LM, Xie DY, Dolezal K, Schlereth A, Jurgens G, Alonso JM (2008) TAA1-mediated auxin biosynthesis is essential for hormone crosstalk and plant development. Cell 133:177–191

Street IH, Mathews DE, Yamburkenko MV, Sorooshzadeh A, John RT, Swarup R, Bennett MJ, Kieber JJ, Schaller GE (2016) Cytokinin acts through the auxin influx carrier AUX1 to regulate cell elongation in the root. Development 143:3982–3993

Tan TH, Silverberg JL, Floss DS, Harrison MJ, Henley CL, Cohen I (2015) How grow-and-switch gravitropism generates root coiling and root waving growth responses in Medicago truncatula. Proc Natl Acad Sci U S A 112:12938–12943

Taylor LP, Grotewold E (2005) Flavonoids as developmental regulators. Curr Opin Plant Biol 8:317–323

Teale W, Palme K (2018) Naphthylphthalamic acid and the mechanism of polar auxin transport. J Exp Bot 69:303–312

Tsuchisaka A, Theologis A (2004) Unique and overlapping expression patterns among the Arabidopsis 1-amino-cyclopropane-1-carboxylate synthase gene family members. Plant Physiol 136:2982–3000

Vandenbussche F, Petrasek J, Zadnikova P, Hoyerova K, Pesek B, Raz V, Swarup R, Bennett M, Zazimalova E, Benkova E, Van Der Straeten D (2010) The auxin influx carriers AUX1 and LAX3 are involved in auxin-ethylene interactions during apical hook development in Arabidopsis thaliana seedlings. Development 137:597–606

Vandenbussche F, Vaseva I, Vissenberg K, Van Der Straeten D (2012) Ethylene in vegetative development: a tale with a riddle. New Phytol 194:895–909

Weerasinghe RR, Swanson SJ, Okada SF, Garrett MB, Kim SY, Stacey G, Boucher RC, Gilroy S, Jones AM (2009) Touch induces ATP release in Arabidopsis roots that is modulated by the heterotrimeric G-protein complex. FEBS Lett 583:2521–2526

Weller B, Zourelidou M, Frank L, Barbosa ICR, Fastner A, Richter S, Juergens G, Hammes UZ, Schwechheimer C (2017) Dynamic PIN-FORMED auxin efflux carrier phosphorylation at the plasma membrane controls auxin efflux-dependent growth. Proc Natl Acad Sci USA 114: E887–E896

Woeste KE, Ye C, Kieber JJ (1999) Two Arabidopsis mutants that overproduce ethylene are affected in the posttranscriptional regulation of 1-aminocyclopropane-1-carboxylic acid synthase. Plant Physiol 119:521–530

Wolverton C, Mullen JL, Ishikawa H, Evans ML (2002) Root gravitropism in response to a signal originating outside of the cap. Planta 215:153–157

Wolverton C, Paya AM, Toska J (2011) Root cap angle and gravitropic response rate are uncoupled in the Arabidopsis pgm-1 mutant. Physiol Plant 141:373–382

Xuan W, Audenaert D, Parizot B, Moller BK, Njo MF, De Rybel B, De Rop G, Van Isterdael G, Mahonen AP, Vanneste S, Beeckman T (2015) Root cap-derived auxin pre-patterns the longitudinal axis of the Arabidopsis root. Curr Biol 25:1381–1388

Zajaczkowska U, Barlow PW (2017) The effect of lunisolar tidal acceleration on stem elongation growth, nutations and leaf movements in peppermint (Menthaxpiperita L.). Plant Biol 19:630–642

Zheng ZY, Zou JJ, Li HH, Xue S, Wang YR, Le J (2015) Microrheological insights into the dynamics of amyloplasts in root gravity-sensing cells. Mol Plant 8:660–663

Zourelidou M, Absmanner B, Weller B, Barbosa ICR, Willige BC, Fastner A, Streit V, Port S, Colcombet J, van Bentem SDLF, Hirt H, Kuster B, Schulze WX, Hammes UZ, Schwechheimer C (2014) Auxin efflux by PIN-FORMED proteins is activated by two different protein kinases, D6 PROTEIN KINASE and PINOID. elife 3

Chapter 8
Bioregenerative Life Support Systems in Space Research

Donat-Peter Häder, Markus Braun, and Ruth Hemmersbach

Abstract For manned long-term missions e.g. to Mars, large amounts of food and oxygen are required to sustain the astronauts during the months- or year-long travel in space but resources are very limited. Water is already routinely recycled on the ISS. In order to solve the problem of limited food and oxygen resources, bioregenerative life support systems are envisioned with closed nutrient and gas loops. Several ecological model systems varying in the degree of complexity have already been investigated on ground and tested on shorter space flights. Photosynthetic organisms such as flagellates or higher plants produce oxygen when light is available. Simultaneously they take up the carbon dioxide exhaled by the astronauts or other consumers. Urea and ammonia can be detoxified by bacteria. Insertion of a component of primary consumers such as ciliates could be used to produce fish for human consumption.

Keywords Life support system · Oxygen · Carbon dioxide · Aquarack · Aquacells · OmegaHab · C.E.B.A.S. · Eu:CROPIS · MELISSA

8.1 Introduction

Manned spaceflight to other planets or extraterrestrial moons require a high level of sophistication. Due to the strong impact of galactic cosmic radiation (composed of high-energy protons and high charge and energy nuclei) the shielding of the spacecraft has to be very strong in order to protect the astronauts during their up to 3 years long journey e.g. to Mars (Cucinotta et al. 2013). With current technology astronauts would have a risk exceeding 5% for mortality and a 10% risk for morbidity, respectively, for a Mars mission. In addition, a large amount of food has to be transported to sustain human travelers, and the status and stability of bioactive compounds have to be warranted for this period. The food can be dehydrated since water is recycled during long missions from wastes and exhaled air to produce potable water (Moores et al. 2015; Barta 2017).

M. Braun et al., *Gravitational Biology I*, SpringerBriefs in Space Life Sciences, https://doi.org/10.1007/978-3-319-93894-3_8

Oxygen sustaining the astronauts during such long-term missions cannot be carried in sufficient quantity. Depending on activity a 75-kg human needs about 700 to 1000 L of oxygen per day (Montoye et al. 1983). Likewise, the exhaled carbon dioxide needs to be removed from the air in addition to nitric oxide and other trace gases (Dweik et al. 1998). NASA is currently using solid amine sorbent and zeolite 5A molecular sieve material packed into beds for removing carbon dioxide during extended space flights (Satyapal et al. 2001; Knox et al. 2015).

The combined problems led to the concept of recycling these materials during extended space flights using bioregenerative life support systems (Blüm 2003; Wang et al. 2006). Based on photosynthetic algae or plants these systems can absorb carbon dioxide, produce oxygen and remove wastes such as ammonia (Yang et al. 1997; Sakano et al. 2002). In addition, advanced ecological life support systems may even be used for crop production for human consumption such as vegetables (Wheeler and Sager 2006).

8.2   Aquarack

In preparation of space flights it was proposed to carry out a long-term experiment growing flagellates in a closed bioreactor (Häder and Kreuzberg 1990). The unicellular green flagellate *Euglena gracilis* was inoculated in autoclaved tap water in a 10-L bioreactor (Braun Biotech, Melsungen, Germany; Porst et al. 1997). The density was adjusted to 1.7×10^5 cells/mL. The reactor was completely closed with no additions of nutrients, and light was the only energy source. Measurements were performed using a closed-loop concept without taking samples. Cell suspension from the bioreactor was pumped through two loops. In one loop the cell suspension was pumped through a vertically oriented, circular viewing chamber which allowed observing the cells under a microscope with an attached CCD camera. Cell density, motility and orientation were analyzed by a real-time image analysis program (Häder 1994). Also in this loop was a flow-through cuvette to perform absorption measurement via a glass-fiber cable connected to a microspectrometer (Ocean Optics Inc., USA). In a second loop the cell suspension was permanently pumped through an electrode holder for on-line water analysis at a flow rate of 15 L/h. Oxygen concentration was determined with a Clark electrode and nitrate with an Orion (Model 83-07) electrode; in addition, the pH was determined on line (Lebert et al. 1995). The signals were amplified and recorded after A/D conversion. Under three fluorescence lamps (mixed cool white and warm tone, 20 W m^{-2}) the flagellates produced about 3.9 mg oxygen per hour which increased to 10.3 mg/h with nine lamps. Because the cells also consumed oxygen by respiration the concentration in the bioreactor was constant at about 8.5 mg/L (Lebert and Häder 1998). The duration of the long-term experiment was more than 600 days.

During the first 5 months the cell density slowly increased up to 3.5×10^5 cell/mL, subsequently decreased to the initial value (1.7×10^5 cells/mL) after 10 months and thereafter was stable for another 11 months. Because no nutrients had been added,

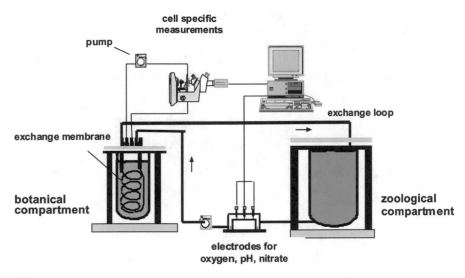

Fig. 8.1 Schematic representation of the Aquarack hardware

there was no pronounced increase in the concentration of cells which were in a stationary phase. The chlorophyll *a* concentration was measured at 675 nm and calculated per cell. The initial concentration decreased first slowly and then at a faster pace. After 540 d it had fallen to 64% of the initial value.

Euglena gracilis shows a pronounced negative gravitaxis (cf. Chap. 3) which was very precise shortly after inoculation as indicated by r-values ≥0.8. The precision of orientation dropped to an r-value of about 0.3 and with increasing culture age decreased to below 0.2 which is regarded to indicate random orientation.

Toward the end of the experiment the bioreactor was coupled with a zoological compartment (60 L aquarium) via a semipermeable membrane tube (2.4 nm pore size) with a surface of 1000 cm^2 floating inside the bioreactor (Porst et al. 1996). The medium of the zoological compartment was pumped through the tube at a rate of 69 L/h and housed four adult swordtail fish (*Xiphophorus helleri*) and 15 snails (*Biomphalaria glabra*; Fig. 8.1). After closing the zoological compartment the oxygen concentration dropped to 4 mg/L and the pH by 1 unit. After activation of the exchange loop the oxygen increased to 7 mg/L, but oscillated due to a higher consumption after feeding the fish. The nitrate concentration was stable around 40 mg/L. After 2 weeks the cellulose exchange membrane had been digested by microorganisms and burst which marked the end of the experiment.

8.3   Aquacells and OmegaHab

Based on these terrestrial experiments several closed environmental life support systems were developed for space experiments on Russian Foton satellites (Fig. 8.2). The *Euglena* suspension was housed in a 1450-mL cylindrical container and

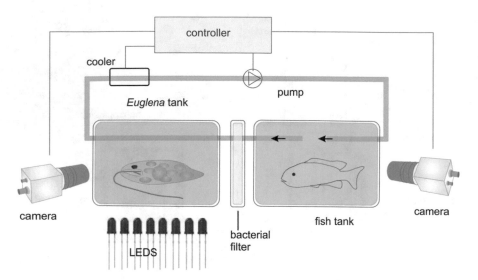

Fig. 8.2 Schematic drawing of the Aquacells hardware

irradiated with red LEDs to sustain photosynthesis. The water from the fish tank (1260 mL with 35 larval cichlids, *Oreochromis mossambicus*) was pumped first through a commercial filter with a bacterial biofilm which converted the ammonium excreted by the fish to nitrate. Subsequently it was transferred through 12 membrane tubes running through the *Euglena* aquarium for exchange of oxygen, carbon dioxide and nitrate. Motility and orientation of both fish and algal cells were recorded at regular intervals on video tape during the mission. The hardware was installed in the Foton satellite and launched on a Soyuz rocket from Baikonur (Russia) for an 11-day mission. As expected, during the space flight the *Euglena* cells swam randomly (r-value 0.03) and at higher velocities than under 1-g conditions in an identical experimental hardware on ground (Häder et al. 2006). After a prolonged time in space the cells took several hours to show again normal gravitaxis, which is in contrast to the short Texus missions, where normal gravitaxis was observed when the cells were returned to the ground. Under microgravity the cells were more rounded than the ground control, indicating that they experienced a pronounced stress condition. The oxygen production was monitored with an Oxy 4 mini system (Precision Sensing GmbH, Regensburg, Germany) and was sufficient to sustain the fish.

A follow-up experiment was launched during the Russian FOTON-M3 mission called OmegaHab (*Oreochromis mossambicus Euglena gracilis* aquatic Habitat (Strauch et al. 2008). Again *Euglena* was the oxygen producer and 26 Tilapia larvae (approximately 12 mm long) the consumers. A hardware device was used to feed the fish periodically. Photosynthesis was sustained by three pairs of high-power red LEDs. During the 12-day orbital flight the behavior of both flagellates and fish larvae was recorded for 10 min every day. During the flight housekeeping data such as temperature were down-linked in order to adjust the ground-based control experiment. Temperature was kept constant by a cooler. All systems worked flaw-lessly, but only 11 fish survived due to an insufficient oxygen concentration toward the end of the experiment. *Oreochromis mossambicus* larvae produced significantly larger utricular otoliths as compared to the 1-g controls (Anken et al. 2016). This result confirmed earlier findings where *Oreochromis* larvae had been kept in a fast rotating, submerged clinostat (Anken et al. 2010). Wall vessel rotation did not impair the growth of cichlid otoliths (Brungs et al. 2011) but it increased the growth of otoliths in zebrafish (Li et al. 2011, 2017a). This difference may be due to the different behaviors of the two species: the cichlids are mouth-breading, whereas the zebrafish lay eggs (Hilbig and Anken 2017).

A more illustrious community was flown in space for 4 weeks with an advanced version of the aquatic OmegaHab system on the Russian Bion-M1 mission in spring 2013 (Hilbig and Anken 2017). OmegaHab was developed into an artificial minia-ture ecosystem consisting of three chambers (about 5 L): one chamber was home to 55 tilapia larvae (*Oreochromis mossambicus*), Mexican fresh water crustaceans (*Hyalella azteca*), a few ram's horn snails (*Biomphalaria glabrata*) and hornwort (*Ceratophyllum demersum*), while a second photobioreactor chamber housed a population of *Euglena gracilis* as primary oxygen producer. Between the chambers a microbial filter populated by nitrifying bacteria that were to decompose the excreta of the fish and transform them into fertilizer for the hornwort and the algae. This closed life cycle mini-ecosystem worked well in microgravity 575 km above the Earth until the LED lighting failed cutting off the "animal crew" from oxygen resupply. While these species did not survive, *Euglena* continued to produce bio-mass by switching from photoautotrophic to heterotrophic mode, feeding on the nutrients released by the decomposing organisms.

In cooperation with Chinese scientists an extremely miniaturized closed aquatic ecosystem of 60 mL total volume was flown on the Shenzhou 8 spacecraft containing *Chlorella*, *Euglena* and *Bulnius* in separate chambers (Li et al. 2017b). One chamber housed the *Euglena* culture, while the other one *Chlorella* cells and three snails. The spacecraft flew in orbit for 17.5 days and one snail survived. The total cell numbers, the assimilation of nitrogen and phosphorous as well as concen-trations of soluble proteins and carbohydrate decreased in the flight module as compared to the ground module.

During the same mission *Euglena* cells were sent to space and fixed 40 min after launch in microgravity with an RNA lysis buffer. In parallel, cells were fixed which had been kept on a 1-g reference centrifuge (Nasir et al. 2014). After return of the samples to ground the transcription of genes involved in signal transduction, oxida-tive stress defense, cell cycle regulation and heat shock responses was analyzed using quantitative PCR. The analysis showed that *Euglena* suffers stress upon short-term exposure to microgravity since of the 32 tested genes, 18 stress-induced genes, involved in signal transduction, oxidative stress defense, cell cycle regulation and heat shock responses, were up-regulated as indicated by quantitative PCR. These results confirm that long-term space flights are valuable tools to study the behavior, physiology and genetics of motile microorganisms which promise further insight into the complex molecular machinery of graviperception, signal transduction and movement control.

8.4   C.E.B.A.S.

In another approach a Closed Equilibrated Biological Aquatic System (C.E.B.A.S.) was developed (Bluem and Paris 2001). This complex system consisted of four components: A zoological component with fishes (*Xiphophorus helleri*) and water snails (*Biomphalaria glabrata*), a microbial compartment with ammonia-oxidizing bacteria, a botanical compartment with the rootless, edible water plant *Ceratophyllum demersum* and an electronic compartment for process control and data acquisition. The underlying idea was that the plants harvest light energy and produce biomass using photosynthesis.

The animals break down the biomass and produce carbon dioxide which is utilized by the plants. The ammonia excreted by the animals is converted by the bacteria to nitrite and subsequently to nitrate which is used by the plants as fertilizer. The 150-L container was tested for more than 13 months. In addition, a C.E.B.A.S. mini module was built (8.5 L). Due to space limitations there was no closed food loop and the animals had to be fed by an automated feeder. This mini module was flown successfully on the STS-89 and STS-90 space shuttle missions. The results indicate that the plant biomass production was not affected by the space conditions and that the immune system of the animals was not disturbed. In a later version the small water plant *Wolffia arrhiza* was utilized which is consumed as vegetable in Southeast Asia. The reproduction of the fish and snails functioned without problems under space conditions for 120 days on the ISS in a mid-deck locker compartment (Blüm 2003). However, harvesting of the various organisms had to be done by the astronauts in order to avoid self-shading of the plants and excessive oxygen consumption and ammonia excretion by the animals. The second problem was that the animals had to be fed manually on a daily basis which demanded too much crew time.

8.5   Eu:CROPIS

A new approach to the development of a regenerative live support system is the planned Eu:CROPIS mission within the DLR compact satellite program (Hauslage et al. 2018). A major problem in manned space flight is the processing of urine. While water is recycled the dissolved substances (urea and salts) are discarded. The new idea is to utilize urine as a fertilizer to grow fruits and vegetables using two life support systems within the compact satellite Eu:CROPIS. This will be accomplished by using a nitrifying trickle filter on a lava rock which houses bacteria, fungi and protozoa. *Euglena* is used to produce oxygen and in addition, this flagellate can convert ammonia to nitrite. Small tomatoes (Micro-Tina) will be used as higher plants. The seeds will be germinated under low-gravity conditions such as 0.16 g (moon) and 0.38 g (Mars) for 6 months each. During the experiment the ion concentration in the liquid flow will be determined by ion chromatography and *Euglena* will be subjected to molecular biology analysis.

8.6 MELISSA

In a regenerative life support system a higher plant compartment was planned for the MIR station to grow vegetables which have the additional advantage of using exhaled carbon dioxide, produce oxygen, reclaim water and provide food for astronauts during extended space travel (Tri et al. 1991). In a European development a loop of interconnected bioreactors has been developed in order to provide life support in space called Micro Ecological Life Support System Alternative (MELISSA). It consists of four bioreactors plus a compartment for higher plants (Godia et al. 2002). One reactor contains a packed bed working with an immobilized culture of *Nitrosomonas* and *Nitrobacter* and an external loop gas-lift photobioreactor for the cyanobacterium *Spirulina platensis*. In order to proof that *Spirulina* can be used as food, groups of five rats were fed for 16 weeks with a diet supplemented with 0–40% dried cyanobacteria as protein source (Tranquille et al. 1994).

In order to improve the stability and safety each compartment has its local control system; in addition, the functioning takes into account the status of the other compartments and the global desired functioning point which determines the setpoints of each compartment based on a non-linear predictive model strategy (Fulget et al. 1999). This system is under the directive of the European Space Agency (ESA) and the MELISSA pilot plant facility has been re-designed and extended as a pilot plant facility and a test-bed for advanced life support systems in order to study the robustness and stability of the continuous operation of a complex biological system (Gòdia et al. 2004). In addition to testing the instrumentation, chemical and microbial safety of the system as well as tracking the genetic stability of the microbial strains was important in order to warrant stable life support systems for long-term manned missions.

The key elements are the production of food and oxygen as well as the regeneration of water from organic wastes including urine and CO_2 (Lasseur et al. 2010). During the first 20 years of development the system has been extended to five compartments ranging from anoxygenic, thermophilic to photo-autotrophic (higher plants).

Thirty organizations from Europe and Canada have signed a memorandum of understanding and the project is managed by ESA. Plants are cultivated by using controlled environment agriculture technologies for space applications developed by the DLR Institute of Space Systems which incorporated the evolution and design of an environmentally-closed nutrition source research group and laboratory (Kolvenbach

2014). One key aspect is the atmosphere management system responsible for providing breathing air with suitable temperature, velocity, humidity and CO_2.

8.7   Outlook

Building on the experience gathered from the C.E.B.A.S, Aquacells/OmegaHab and MELISSA projects, the next generation of bioregenerative systems is now being developed within European space life sciences programs in a modular approach focusing on dissecting the complex interactions and interdependencies between two and more species in a bioregenerative closed ecological life support system. Maximizing the efficiency of a single component like a photobioreactor based on understanding the precise needs of a photosynthetic organism still is a surprisingly big challenge today but developing bioregenerative life support systems that may be used to complement physico-chemical systems or even to provide food for long-term space travelers in a closed-loop manner will be an essential precondition for the future of humans exploring the solar system and the universe.

References

Anken RH, Baur U, Hilbig R (2010) Clinorotation increases the growth of utricular otoliths of developing cichlid fish. Microgravity Sci Technol 22:151–154

Anken R, Brungs S, Grimm D, Knie M, Hilbig R (2016) Fish inner ear otolith growth under real microgravity (spaceflight) and clinorotation. Microgravity Sci Technol 28:351–356

Barta DJ (2017) Getting out of orbit: water recycling requirements and technology needs for long duration missions away from earth

Bluem V, Paris F (2001) Aquatic modules for bioregenerative life support systems based on the C.E.B.A.S. biotechnology. Acta Astronaut 48:287–297

Blüm V (2003) Aquatic modules for bioregenerative life support systems: developmental aspects based on the space flight results of the C.E.B.A.S. mini-module. Adv Space Res 31:1683–1691

Brungs S, Hauslage J, Hilbig R, Hemmersbach R, Anken R (2011) Effects of simulated weight-lessness on fish otolith growth: clinostat versus rotating-wall vessel. Adv Space Res 48:792–798

Cucinotta FA, Kim M-HY, Chappell LJ, Huff JL (2013) How safe is safe enough? Radiation risk for a human mission to Mars. PLoS One 8:e74988

Dweik RA, Laskowski D, Abu-Soud HM, Kaneko F, Hutte R, Stuehr DJ, Erzurum SC (1998) Nitric oxide synthesis in the lung. Regulation by oxygen through a kinetic mechanism. J Clin Investig 101:660

Fulget N, Poughon L, Richalet J, Lasseur C (1999) MELISSA: global control strategy of the artificial ecosystem by using first principles models of the compartments. Adv Space Res 24:397–405

Godia F, Albiol J, Montesinos J, Pérez J, Creus N, Cabello F, Mengual X, Montras A, Lasseur C (2002) MELISSA: a loop of interconnected bioreactors to develop life support in space. J Biotechnol 99:319–330

Gòdia F, Albiol J, Pérez J, Creus N, Cabello F, Montras A, Masot A, Lasseur C (2004) The MELISSA pilotplant facility as an integrated test-bed for advanced life support systems. Adv Space Res 34:1483–1493

Häder D-P (1994) Real-time tracking of microorganisms. Binary 6:81–86

Häder D-P, Kreuzberg K (1990) Algal bioreactor-concept and experiment design. Proceedings of the workshop (DARA/CNES) on artificial ecological systems, 24–26 October 1990, Marseille

Häder D-P, Richter PR, Strauch SM, Schuster M (2006) Aquacells – flagellates under long-term microgravity and potential usage for life support systems. Microgravity Sci Technol 18:210–214

Hauslage J, Strauch SM, Eßmann O, Haag FWM, Richter P, Krüger J, Julia Stoltze J, Becker I, Adeel Nasir A, Bornemann G, Müller H, Delovski T, Berger T, Rutczynska A, Lebert M (2018) Eu:CROPIS – Euglena combined regenerative organic-food production in space. A compact satellite mission testing biological life support systems under lunar and Martian gravity

Hilbig R, Anken R (2017) Impact of micro-and hypergravity on neurovestibular issues of fish. In: Hilbig R, Gollhofer A, Bock O, Manzey D (eds) Sensory motor and behavioral research in space. Springer, Heidelberg, pp 59–86

Knox JC, Gauto H, Miller LA (2015) Development of a test for evaluation of the hydrothermal stability of sorbents used in closed-loop CO_2 removal systems. 45th international conference on environmental systems, 12–16 July 2015. Bellevue, WA

Kolvenbach H (2014) Development of an atmosphere management system for bio-regenerative life support systems. RWTH, Aachen

Lasseur C, Brunet J, De Weever H, Dixon M, Dussap G, Godia F, Leys N, Mergeay M, Van Der Straeten D (2010) MELISSA: the European project of closed life support system. Gravit Space Res 23:3–12

Lebert M, Häder D-P (1998) Aquarack: long-term growth facility for 'professional' gravisensing cells. Proceedings of the 2nd European symposium on the utilisation of the international space station, ESTEC, Noordwijk, The Netherlands. 16–18 November 1998 (ESA-SP 433)

Lebert M, Porst M, Häder D-P (1995) Long-term culture of *Euglena gracilis*: an AQUARACK progress report. Proceedings of the 11th C.E.B.A.S. workshops. Annual issue 1995, Ruhr-University of Bochum

Li X, Anken RH, Wang G, Hilbig R, Liu Y (2011) Effects of wall vessel rotation on the growth of larval zebrafish inner ear otoliths. Microgravity Sci Technol 23:13–18

Li X, Anken R, Liu L, Wang G, Liu Y (2017a) Effects of simulated microgravity on otolith growth of larval zebrafish using a rotating-wall vessel: appropriate rotation speed and fish developmental stage. Microgravity Sci Technol 29:1–8

Li X, Richter PR, Hao Z, An Y, Wang G, Li D, Liu Y, Strauch SM, Schuster M, Haag FW (2017b) Operation of an enclosed aquatic ecosystem in the Shenzhou-8 mission. Acta Astronaut 134:17–22

Montoye HJ, Washburn R, Servais S, Ertl A, Webster JG, Nagle FJ (1983) Estimation of energy expenditure by a portable accelerometer. Med Sci Sports Exerc 15:403–407

Moores JE, Lemmon MT, Rafkin SC, Francis R, Pla-Garcia J, de la Torre Juárez M, Bean K, Kass D, Haberle R, Newman C (2015) Atmospheric movies acquired at the Mars science laboratory landing site: cloud morphology, frequency and significance to the gale crater water cycle and phoenix mission results. Adv Space Res 55:2217–2238

Nasir A, Strauch S, Becker I, Sperling A, Schuster M, Richter P, Weißkopf M, Ntefidou M, Daiker V, An Y (2014) The influence of microgravity on *Euglena gracilis* as studied on Shenzhou 8. Plant Biol 16:113–119

Porst M, Lebert M, Häder D-P (1996) Long-term culture of *Euglena gracilis*: an Aquarack progress report. In: Proceedings of the 11th C.E.B.A.S. Workshops. Ruhr-University, Bochum, pp 217–223

Porst M, Lebert M, Häder D-P (1997) Long-term cultivation of the flagellate *Euglena gracilis*. Microgravity Sci Technol 10:166–169

Sakano Y, Pickering KD, Strom PF, Kerkhof LJ (2002) Spatial distribution of total, ammonia-oxidizing, and denitrifying bacteria in biological wastewater treatment reactors for bioregenerative life support. Appl Environ Microbiol 68:2285–2293

Satyapal S, Filburn T, Trela J, Strange J (2001) Performance and properties of a solid amine sorbent for carbon dioxide removal in space life support applications. Energy Fuel 15:250–255

Strauch S, Schuster M, Lebert M, Richter P, Schmittnagel M, Hader D-P (2008) A closed ecological system in a space experiment. Life in space for life on earth. ESA, Angers

Tranquille N, Emeis J, De Chambure D, Binot R, Tamponnet C (1994) *Spirulina* acceptability trials in rats. A study for the "Melissa" life-support system. Adv Space Res 14:167–170

Tri TO, Brown MF, Ewert MK, Foerg SL, McKinley MK (1991) Regenerative life support systems (RLSS) test bed development at NASA-Johnson Space Center, SAE technical paper

Wang G, Chen H, Li G, Chen L, Li D, Hu C, Chen K, Liu Y (2006) Population growth and physiological characteristics of microalgae in a miniaturized bioreactor during space flight. Acta Astronaut 58:264–269

Wheeler RM, Sager JC (2006) Crop production for advanced life support systems. Technical Reports: 1

Yang VC, Bartlett RH, Palsson BO, Javanmardian M (1997) Photobioreactors and closed ecological life support systems and artificial lungs containing the same, Google Patents

Printed in the United States
By Bookmasters